SpringerBriefs in Economics

For further volumes:
http://www.springer.com/series/8876

Carol Yeh-Yun Lin • Leif Edvinsson
Jeffrey Chen • Tord Beding

National Intellectual Capital and the Financial Crisis in Greece, Italy, Portugal, and Spain

Springer

Carol Yeh-Yun Lin
Department of Business Administration
National Chengchi University
Taipei, Taiwan

Jeffrey Chen
Accenture
Chicago, IL, USA

Leif Edvinsson
Universal Networking Intellectual Capital
Norrtälje, Sweden

Tord Beding
TC-Growth AB
Gothenburg, Sweden

ISSN 2191-5504 ISSN 2191-5512 (electronic)
ISBN 978-1-4614-5989-7 ISBN 978-1-4614-5990-3 (eBook)
DOI 10.1007/978-1-4614-5990-3
Springer New York Heidelberg Dordrecht London

Library of Congress Control Number: 2012951040

Printed on acid-free paper

Springer is part of Springer Science+Business Media (www.springer.com)

Foreword I

The economic crisis is a consequence of many parallel factors which are all related to globalization and digitalization. My main concern, assessing this in more detail from the European perspective, is that revolutionary global forces have not been taken early nor seriously enough by most national and regional decision makers. The Heads of European States and Governments have once again recalled the importance of fiscal consolidation, structural reform and targeted investment to put Europe back on the path of smart, sustainable and inclusive growth. The main question is how capable and ready are the national governments to tackling the complex and manifold issues of crises and to renewing even radically many of our public and private structures and processes.

The first basic requirement is that all the European Union Member States remain fully committed to taking the actions required at the national level to achieve the objectives of the Europe 2020 Strategy. The second basic requirement is that the national and regional governments, as well as people, are ready for radical changes. This booklet, and the other 11 booklets by the experienced authors, focus on national intellectual capital and give necessary insights and facts for us the readers and especially for our in-depth systemic thinking of the interrelationships of NIC and economic recovery.

How should the national and regional decision makers tackle the existing knowledge of intangible capital? The focus needs to be more on the bottom-up approach stressing the developments on local and regional levels. I highlight our recent statements by the EU Committee of the Regions. The key priorities are to get more innovations out of research and to encourage mindset change towards open innovation.

The political decision makers are finally aware that the traditional indicators created for and used in industrial production cannot be applied to a knowledge-intensive, turbulent and innovativeness-based global enterprise environment. Indicators that perceive the intangible dimensions of competitiveness – knowledge capital, innovation knowledge and anticipation of the future – have been developed around the world, but their use has not yet become established in practice. This booklet accelerates the development and the use of these indicators.

This helps the local and regional, as well as central, governments in taking brave leaps forward on a practical level – giving greater ownership and involving all the stakeholders. This means the need of actions towards increasing the structural and relational capital of regions, both internally in communities of practice and in collaboration with others.

The new generation innovation activities are socially motivated, open and collectively participated, complex and global by nature. The regions need to move towards open innovation, within a human-centered vision of partnerships between public and private sector actors, with universities playing a crucial role.

Regions should be encouraged to develop regional innovation platforms, which act as demand-based service centres and promote the use of international knowledge to implement the Europe 2020 Strategy, smart specialization and European partnerships according to the interests and needs of regions. For this to happen, we need to apply the new dynamic understanding of regional innovation ecosystems, in which companies, cities and universities as well as other public and private sector actors (the "Triple Helix") learn to work together in new and creative ways to fully harness their innovative potential.

New innovative practices do not come about by themselves. One major potential is the use of public procurement. The renewing of the European wide rules must increase the strategic agility and activities of municipalities and other public operators as creators of new solutions. Especially the execution of pre-commercial procurement should be reinforced even more in combination with open innovation to speed up the green knowledge society development, i.e. for common re-usable solutions in creating the infrastructures and services modern real-world innovation ecosystems are built upon. Conditions must be created that also allow for extensive development projects which address complex societal challenges and which take the form of risk-taking consortia.

One of our working instruments within the Committee of the Regions is the Europe 2020 Monitoring Platform, which broadly reviews and reflects the opinions and decisions on regional level all around Europe. It gives a flavor of cultural and other socio-economic differences inside the EU. This brings an important perspective to the intellectual capital, namely the values and attitudes needed for citizens supporting policymakers on appropriate long-term investments and policies.

Emphasizing the importance of these issues, decision-makers in all countries and regions worldwide need a deep and broad understanding of the critical success factors affecting the national intellectual capital. With all the facts and frames for thinking this booklet gives a valuable insight in today's challenges.

Markku Markkula
Advisor to the Aalto University Presidents
Member of the EU Committee of the Regions
Former Member of the Parliament of Finland

Foreword II

Longitudinality is the key.

Whether it is on the sporting field or in global markets. Understanding the nuances of how performance changes over time is critical for appreciating the true nature of competitiveness. A world class footballer can effectively evaluate the tendencies of his opponent as the game progresses. Does he favour turning left or right? Will he approach the near or far post when targeting his run towards the goal? Over time, patterns emerge and the top goal scorers exploit this knowledge for competitive gain. The early portion of the match is used as a predictor of the behaviours that will be exhibited later on.

Global markets are no different. As national political parties in power ebb and flow, economic indicators adjust to reflect the markets. Naturally, an early investment in education will yield a higher corresponding literacy rate. Logically, a large investment in telecommunications infrastructure will yield higher internet penetration rates. The main difference between international markets and a football match, is the temporal lag, or longitudinality.

In this insightful booklet, Drs. Carol Lin and Leif Edvinsson take the reader on a journey of longitudinality. The setting happens to be the economic crisis of Greece, Italy, Portugal and Spain. Against this backdrop of four Southern European nations, Lin and Edvinsson weave a masterful collection of insights and metrics to determine whether or not the economic crisis could have been predicted. Indeed, this is a critically important research program with enormous implications for economies that go way beyond European borders.

The premise is simple yet powerful. Can national intellectual capital indicators yield a warning for pending economic crisis? In this booklet, the hypothesis is thoroughly tested and validated. The national intellectual capital literature has its genesis with the transformation of the traditional intellectual capital framework (and its corresponding firm-level perspective) into a country-level point of view. Various researchers have examined the adequacy of the framework at a national level and have found it to be robust. In fact, several empirical studies including many of my own have shown statistically significant linkages between various inputs (e.g., human capital, process capital, renewal capital) and corresponding traditional financial outcomes (e.g., GDP per capita PPP).

However, this particular research program fills a very important void in the investigative landscape because its ultimate goal is to provide strategists, policy makers and analysts with a stronger set of tools for determining the competitiveness of nations. Whereas traditional economic measures have been evident for decades, this novel approach boasts a more meaningful and accurate assessment of what has transpired in Southern Europe recently.

Of course, the impact of Lin and Edvinsson's work will be evaluated against the test of time … as it should be. The longitudinal nature of a nation's economic ups and downs is not an easy formula to crack. If it was, the global economy would not be in the position it is in. As an academic researcher, I am keenly excited about the results of this study and the new directions this research program will lead to. As a management consultant, I am eager to comprehend exactly what policies governments can embrace to lift themselves out of a pending economic crisis. Most importantly, as a proud Greek-Canadian, I am extremely worried that the cradle of civilization, philosophy and democracy that was born thousands of years ago, could slowly erode. This would be very sad for a proud nation and its heritage. While I don't condone some of the financial irresponsibilities that have transpired recently in Greece, I will always be proud of my forefathers and will continue to enjoy the sun-bathed beaches that I proudly hail as my homeland. Perhaps this booklet will act as an inflection point in Southern Europe's own longitudinality.

Dr. Nick Bontis
Director, Institute for Intellectual Capital Research
McMaster University, Canada
Kryopigi, Halkidiki, Greece
www.NickBontis.com

Foreword III

The 2008 global financial crisis hit the whole world with unprecedented speed, causing widespread financial panic. Consumer confidence dropped to the lowest level since the Great Depression. Taiwan, with an export-dependent economy, was seriously impacted by the crisis and the unemployment rate hiked while household consumption levels dropped. At the onset of the financial crisis, Professor Lin was the Dean of Student Affairs here at National Chengchi University in Taipei, Taiwan. She was the dean in charge of financial aid and student loans and thus saw firsthand the direct impact the financial crisis had upon our students. The crisis was so devastating that Professor Lin, along with the university, was compelled to launch several new initiatives to raise money and help students weather the difficult times.

I am very glad that she took this painful experience to heart and set herself upon the task of investigating the impact of the crisis; trying to look into the causes and consequences for policy implications, not only for Taiwan but for an array of 48 countries. In particular, she approaches the crisis from the perspective of "national intellectual capital," which is very important in today's knowledge-driven economy.

Taiwan is an example of a knowledge economy and has enjoyed the fame of being referred to as a "high-tech island." Without an abundance of natural resources, Taiwan's hardworking and highly-educated population is the single most precious resource that the island has. Acknowledging the value of such human resources and intellectual capital, we established the Taiwan Intellectual Capital Research Center (TICRC) under my leadership in 2003. Ever since then, Taiwan's government has continuously funded the university to conduct relevant research projects aimed at enhancing the intellectual capital of Taiwan. Having been thus endowed with the responsibility of nourishing future leaders in the public and private sectors, we have focused on building up our strength in innovation, entrepreneurship, and technology management related research and education.

To enhance intellectual capital research, we recently formed a joint team of professors for a four-year project in order to leverage their respective research capabilities. Through this project we hope to provide policy suggestions for the government by exploring the creativity, innovation and intellectual capital at national, regional, city and county levels. The goal is to come up with an intangible assets agenda for Taiwan's future sustainability. Professor Lin is an integral member in this research team.

Following her 2011 book *National Intellectual Capital: A Comparison of 40 Countries*, this booklet series is Professor Lin's second attempt at presenting her research, conducted under the sponsorship of TICRC, to international readers. As the Founding Director of TICRC and her President, I am honored to give a brief introduction of the value of this booklet series.

In comparison to her 2011 book, this series increased the number of countries studied to 48 and particularly focuses on the impact of intellectual capital on the 2008 global financial crisis. Rarely has an economic issue been systematically studied from the view point of intangible assets, particularly at such a large scale of 48 countries. The research results show without a doubt that national intellectual capital is indeed an important economic development enhancer. In particular, the fact that countries with higher national intellectual capital experienced faster recoveries from the 2008 financial crisis provides a strong message for the policy makers.

In addition to providing insights to national policy, the booklet also summarizes the background of each country before the crisis, the key events during the crisis, economic development afterwards, and future prospects and challenges. Each volume affords readers a holistic picture of what happened in each country in an efficient manner. The linkage between national intellectual capital and this financial crisis also provides a different perspective of the crisis.

We are happy that Professor Lin continues to share her valuable research results with international readers. I sincerely hope that her insights can garner more attention concerning the benefits of developing national intellectual capital for the well-being of every nation.

Se-Hwa Wu
Professor, Graduate Institute of Technology
and Innovation Management
President, National Chengchi Univeristy,
Taipei, Taiwan

Preface I

There are "mounting risks of a breakup of the Euro zone." Such comments are frequent today on how the European leaders are handling the escalating crisis and its potential impact on non-European countries. But few leaders, reporters or researchers are actually addressing the situation of national intellectual capital (NIC) and its signals. In addition to the financial crisis, is there an emerging NIC crisis as well? Why is it emerging? How should policy makers think about NIC? In what way does it need specific attention? When will the outcome and impact of taken NIC policy steps be realized?

In the midst of the European crisis, there are national interventions to address the issues mentioned above. In leading economical nations the investments going into intangibles now exceeds tangibles, and is positively correlated to income per capita. However, these still do not show up clearly in national mapping as well as policy making insights. Therefore the New Club of Paris is focusing the knowledge agenda setting for countries on Societal Innovation (see www.new-club-of-paris.org).

Chairman Ben Bernanke of the U.S. Federal Reserve was addressing some of these same aspects in a key note speech in May 2011 hosted by Georgetown University: http://www.icapitaladvisors.com/2011/05/31/bernanke-on-intangible-capital/. OECD and the World Bank are developing NIC statistics, often based on the model from Corrado-Hultén. Japan has been developing both NIC and Intangible Assets (IA) at METI for some time now. Their research on IC/IA has resulted in a National IA Week with various key stakeholders, such as government agencies, universities, stock exchange and enterprises. Japan is so far the only country in the world to hold such activities, and they have been doing so for the last eight years. Australia, Singapore, South Korea, and China are currently undertaking various NIC initiatives. Other countries are also becoming more and more aware of NIC, with policy rhetoric centered on innovation, education, R&D, and trade. Despite this, the map for a more justified NIC navigation has been missing.

This booklet highlights NIC development for a number of countries, based on 28 different indicators, aggregated into four major NIC components of human capital, market capital, process capital and renewal capital. The model here is a refined and verified statistical model in comparison to the Corrado-Hultén model. We call it the

L-E-S model after the contributors Lin-Edvinsson-Stahle. Based on a deeper understanding and the timeline pattern it sets forth, this model will add to a better NIC navigation, not to mention knowledge agenda setting for countries.

Upon looking at a global cluster NIC map, it is evident that the top leading countries seem to be small countries, especially Singapore, the Nordic countries, Hong Kong and Taiwan. For the U.S., Finland and Sweden around 50 % or more of its economical growth is related to NIC aspects. Sweden, Finland, Switzerland, the U.S., Israel and Denmark are strongly influenced in its GDP growth by focusing on Renewal Capital.

It might be that we will see a clearer map of the NIC ecosystem and drivers for wealth emerge in the extension of this ongoing unique research of NIC. This booklet will present a NIC map for various clusters of countries. It can be used for bench marking as well as bench learning for policy prototyping. The starting point is awareness and thinking of NIC, and its drivers for economic results. Based on this more refined navigation, NIC metrics can be presented.

Deeper understanding will emerge from this research, such as the scaling up of limited skilled human capital in one nation by using the globalized broadband technologies for migration and flow of knowledge (such as tele-medicine or mobile banking in Africa). This is also referred to as the IC multiplier. It might also be the way the old British Commonwealth was constructed, but without the IC taxonomy. In modern taxonomy it might be the shaping of NIC alliances for the migration and flow of IC between nations?

Another understanding that might emerge for policy making is the issue of employment versus unemployment. The critical understanding will be deployment of IC drivers. This will require another networked workforce of value networkers on a global scale, such as volunteering software and apps developers. However such volunteers do not show up in traditional statistics, for the mapping on behalf of policymakers.

On another level there might be a clear gap analyses between nations to support the vision process of a nation. On a deeper level it is also a leadership responsibility to address the gap of NIC positions versus potential positions. Such a gap is in fact a liability to the citizens, to be addressed in due time.

This will take us to the need for the continuous renewal of social systems. The so called Arab Spring is explained by some as resulting from three drivers: lack of renewal of social systems, Internet, and soccer as cross class interaction space. The lack of social renewal and innovation is most likely critical early warning signals. For Greece, we can see such a tipping point occurred back in 1999.

On a global scale we might see that the concern for the Euro zone crisis should and can be explained by a deeper and supplementary understanding of National Intellectual Capital, in addition to financial capital. So we need to refine our NIC understanding, NIC mapping, NIC metrics and NIC organizational constructs into societal innovation for the benefit of wealth creation of subsequent generations.

Leif Edvinsson
The World´s First Professor of Intellectual Capital
Chairman and Co-founder of New Club of Paris

Preface II

Our first book *National Intellectual Capital: A Comparison of 40 Countries* was published in early 2011, at a time when the 2008 global financial crisis had been declared over yet the European region was still plagued with sovereign debt problems. Before we finalized the book, we were able to retrieve some of our raw data concerning the troubled countries, such as Greece, Iceland, Ireland, Portugal, and Spain. The results of our analysis based on data spanning 1995 to 2008 revealed some early warning signs of the financial turmoil in those countries. In my preface of that book, I mentioned the warning signs might reveal only the tip of an iceberg. At that time, my co-author, Professor Edvinsson and I decided to do a follow up study to trace the development of national intellectual capital (NIC) in as many countries as possible, particularly through the lens of the 2008 global financial crisis. This 12 booklet series is the result of that determination.

The 2008 global financial crisis came with unexpected speed and had such a wide-spread effect that surprised many countries far from the epicenter of the initial U.S. sub-prime financial problem, geographically and financially. According to reports, no country was immune from the impact of this financial crisis. Such development clearly signifies how closely connected the world has become and the importance of having a global interdependent view. By reporting what happened during 2005–2010 in 48 major countries throughout the world, this booklet series serves the purpose of uncovering national problems before the crisis, government coping strategies, stimulus plans, potential prospects and challenges of each individual country, and the interdependence between countries. The six years of data allow us to compare NIC and economic development crossing before, during, and after the financial crisis. They are handy booklets for readers to have a quick yet overall view of countries of personal interest. The list of 48 countries in 11 clusters is provided in the appendix of each booklet.

Searching for financial crisis related literature for 48 countries is itself a very daunting task, not to mention summarizing and analyzing it. For financial crisis related literature, we mainly relied on the reports and statistics of certain world organizations, including OECD, World Bank, United Nations, International Monetary Fund (IMF), European Commission Office, the US Congressional

Research Service, the U.S. Central Intelligence Agency, and International Labor Office (ILO). Some reliable research centers, such as the National Bureau of Economic Research in the U.S., World Economic Forum, the Heritage Foundation in the U.S., and government websites from each country were also our sources of information. Due to the requirement of more update and comprehensive information, we were not able to use as much academic literature as we would have liked, because it generally covers a very specific topic with time lag and with research methods not easily comprehended by the general public. Therefore, we had to resort to some online news reports for more current information.

In the middle of 2012, the lasting financial troubles caused the European economy to tilt back into a recession, which also slowed down economic growth across the globe. However almost four years have passed since the outbreak of the global financial crisis in late 2008; it is about time to reflect on what happened and the impact of the financial crisis. By comparing so many countries, we came to a preliminary conclusion that countries with faster recovery from the financial crisis have higher national intellectual capital than those with slower recovery. In other words, countries that rebounded fast from the crisis generally have solid NIC fundamentals, including human capital, market capital, process capital, and renewal capital. We also found that the higher the NIC, the higher the GDP per capita (ppp). This booklet series provides a different perspective to look beyond the traditional economic indicators for national development.

In an era when intangible assets have become a key competitive advantage, investing in national intellectual capital development is investing in future national development and well-being.

Enjoy!

Carol Yeh-Yun Lin
Professor, Dept. of Business Administration
National Chengchi University, Taiwan
Taiwan Intellectual Capital Research Center (TICRC)

Executive Summary

Economies with faster recovery from the financial crisis have higher national intellectual capital than those with slower recovery.

How can national intellectual capital (NIC) be supportive as an early warning or policy guideline for national well-being? One key factor in the financial crisis was that the conventional financial system failed to detect potential risks due to non-transparent information disclosure. Our earlier NIC research has revealed certain warning signs of impending financial crisis for Greece, Iceland and Ireland. Such findings indicate that NIC, albeit intangible, can provide valuable insights into future risk control and strategy formulation. This booklet looks into the connections between the financial crisis and NIC development.

Based on data covering 2005–2010 for 48 countries, the figures and tables presented in this booklet largely reflect situations in the real economy, with some statistics showing early warnings before the financial crisis. After the financial crisis, the intangible NIC development also depicted Greece as the poorest performer, then followed by Portugal, with Italy and Spain together rounding up the rear. Data of 48 countries indicate that the higher the NIC, the higher the GDP per capita (ppp), accentuating the value of NIC in major countries throughout the world. **For the NIC ranking, Portugal ranks #26, Spain #27, Italy #28, and Greece #31.**

The 2008 financial crisis is considered to be the worst since the Great Depression of the 1930s, with severe impacts being felt all across the globe. This financial crisis came with an unexpected speed and spread into a global economic shock, resulting in a number of European bank failures. During this period, economies worldwide slowed, credits tightened, and international trade declined. Governments and central banks worldwide responded to the crisis with unprecedented fiscal stimuli, monetary policy expansions and institutional bailouts in their respective countries. While the financial crisis was declared over in the end of 2009, economic recovery in most of the developed countries continues to trudge along beneath pre-recession levels and the Euro zone is still troubled by sovereign debt problems.

At the initial stage of this global financial crisis, the four Southern European countries fared relatively well. Yet when global credits tightened and export demands drastically dropped, their chronic high government deficit and heavy debt failed to withstand the impact. The most serious sovereign debt crisis was set off in late 2009 in Greece when its government admitted that its deficit would be 12.7 % of its gross domestic product, not the 3.7 % the government had forecasted earlier. In early 2010, fears over a potential default grew into a full-fledged financial panic. As the problem spread, Greece secured a bailout package worth US$158 billion (€110 billion) in May 2010. One year later, Portugal was also in deep trouble and obtained an international aid package worth US$110 billion (€78 billion). In 2011, all four countries were known for their faltering economies and burgeoning national debt. Fearing the domino effect, EU leaders had been trying very hard to prevent Italy and Spain from needing a bailout.

As sovereign debt problems continued to ail the Euro zone, major economies were tilting back into recession in late 2011. In March 2012, European finance ministers approved a second bailout of US$172 billion (€130 billion) for Greece. As of mid-2012, the returning recession has halted recovery from the 2008 global financial crisis in some European countries. Despite diligent effort by France and Germany to prevent Italy and Spain from needing a bailout, the Spanish government still requested European financing mainly to recapitalize its failed banks. The Euro zone agreed to lend Spain up to US$125 billion (€100 billion) in financial assistance on June 10, 2012. All four countries continue their structural reforms and consolidation plans, aiming for a more resilient economy and to reduce government debt and deficit.

In general, the NIC of these four countries are in the middle range among the 48 countries with relatively small progress being gained over the six year period. The two short-term oriented NIC – market capital and process capital – started to decline from 2006 and 2007 in Greece and Portugal. Greece in particular has been losing international competitiveness from market capital and process capital after the financial crisis. In addition, the renewal capital of the four countries was grouped together with developing countries even though they are Stage 3 innovation-driven economies according to the World Economic Forum categorization.

The 3D trajectory analysis reveals that **government related issues are the greatest barricade** to achieve GDP growth. The first five areas that need further enhancement are quite similar for these four countries, including three (image of country, transparency of government policy, and fair business competition environment) that require government reforms and two (employee training and computers per capita) that need resource input. Based on deficiency scores, Greece has the most amount of work to do in order to achieve the targeted GDP, followed by Italy, Portugal, and Spain.

This economic crisis provides an ideal opportunity for nations to examine the soundness of their economic system and the effectiveness of national governance related to NIC. The following implications are drawn from our research findings. Readers can refer to Chapter 5 for the rationale behind these implications.

1. National intellectual capital development goes together with economic development and should be regarded as an enhancer of economic growth.
2. More effective national and NIC governance system need to be established.

3. National statistics need to be carefully studied, refined for NIC purposes, effectively interpreted, and strategically utilized for NIC policy.
4. Educating public officials on effectively managing public assets for the benefits of the whole country is essential.
5. National pride and trust should be cultivated within the citizens for community and social capital platform evolution.
6. Utilizing and in-sourcing human capital to revitalize and energize economies may be a plausible measure.
7. Governments need to aggressively launch structural reforms for a more business-friendly and value-creating internal, as well as external, environment in order to boost future economic development and well-being.
8. Economies with faster recovery from the financial crisis have higher national intellectual capital than those with slower recovery.

The data and development of market capital and process capital reported in this booklet clearly reveals early warning signs for Greece and Portugal as early as 2006 and 2007, at a time when they still had good economic growth. If the warnings were picked up early enough, these two nations could have been more resilient to this financial crisis. Therefore our NIC intelligence suggests, in an era when the intangible asset has become a key competitive advantage, investing in national intellectual capital development is investing in future economic development and well-being. National intellectual capital evolution can be nourished both from local culture viewpoint as well as global interconnectivity by social media.

Based on the emerging new insights of values, societal history as well as citizen relationships, a key focus for the future will be on the fusion of national intellectual capital and social service innovation as well as societal innovation, for the enabling of a new societal fabric.

Contents

List of Figures

List of Tables

Appendices

Chapter 1
Introduction

The 2008 global financial crisis is considered by many economists to be the worst one since the Great Depression of the 1930s. The crisis rapidly developed and spread into a global economic shock, which resulted in a number of European bank failures (Fackler 2008; Altman 2009). World political leaders, national ministers of finance, and central bank directors coordinated their efforts to reduce fear, but the crisis continued and eventually led to a global currency crisis. During this period, economies worldwide slowed, credits tightened, and international trade declined.

After the full force of the financial crisis hit in October 2008, there was a call in the United States for a hands-off policy in order to let the markets "work themselves out," in accordance with how capitalism was theoretically supposed to work. Yet, US Federal Reserve Chairman Ben Bernanke believed that such a policy would be catastrophic and urged government intervention. In a statement to congress, Mr. Bernanke said, "If we let the banking system fail, no one will talk about the Great Depression anymore, because this will be so much worse" (Reavis 2009). As a result of this, the US government decided to take action to prevent such a failure.

Following the lead of the United States, governments and central banks worldwide responded to the international crisis with unprecedented fiscal stimuli, monetary policy expansions, and institutional bailouts in their respective countries. The financial rescue worked and an economic crisis akin to the Great Depression was avoided. In fact, the crisis was declared over by the third quarter of 2009 (Kehoe 2010). Yet while the crisis was officially over, economic recovery in most developed countries continued to trudge along beneath prerecession levels with governments continuing to struggle with their finances (Norris 2011).

During the early stage of the financial crisis, management scholars criticized traditional accounting system's inability to reveal the intangibles that explain hidden risks as well as values for proper decision making (Reavis 2009). In line with the importance of the intangibles, intellectual capital (Edvinsson and Malone 1997) advocating the value of human capital, social capital, and the like has gained increasing attention in today's keener global competition.

C.Y.-Y. Lin et al., *National Intellectual Capital and the Financial Crisis in Greece, Italy, Portugal, and Spain*, SpringerBriefs in Economics 7,
DOI 10.1007/978-1-4614-5990-3_1, © The Author(s) 2013

It is our deep belief that national intellectual capital, albeit intangible, can provide valuable insights to policy makers regarding future risk control and strategy formulation. Our previous book, *National Intellectual Capital: A Comparison of 40 Countries* (Lin and Edvinsson 2011; www.nic40.org), was born out of this belief and traces the national intellectual capital development of 40 countries over a 14-year period from 1995 to 2008. The data analysis presented in our previous work revealed certain warning signs of impending financial crisis for countries such as Greece, Iceland, and Ireland (Lin and Edvinsson 2011; 327–333). As a follow-up study, this booklet series is an attempt to further understand the connections between the 2008 global financial crisis and national intellectual capital (NIC) development.

The booklet series, in its entirety, will examine the NIC statuses of 48 countries from the period of 2005 to 2010 to glean new understanding about whether there is a NIC development pattern that distinguishes fast recovery countries from slow recovery ones. This is presented through a series of 11 country clusters, with one booklet focusing on one particular cluster. The clusters are determined based on several factors: geographical proximity, geographical size, or phase of economic development. Focusing on one cluster at a time, we probe the areas of concern within a single country and extend them to compare multiple countries to see whether the situation before and after the crisis can be explained by the intangible NIC. Our data comes from the well-recognized International Institute for Management Development (IMD) in Switzerland. The IMD has been publishing yearly rankings of World Competitiveness for around two decades. Hopefully, the analysis in this booklet series can provide a different ex post perspective of the financial crisis for future policy implications.

This volume—Volume One—will focus on the four Southern European countries of Greece, Italy, Portugal, and Spain as they have economic, market, and inherent similarities that make it easy to compare and examine them as a cluster. First, this booklet will provide an economic background to these four countries as a whole before going into each individual country's development. Through this process, the authors hope to paint a general picture of the economic condition and provide a basis for our dataset and analysis in future sections.

Next, Chap. 2 briefly introduces the impact of 2008 financial crisis on the four countries. Chapter 3 discusses the NIC development of these four countries. Chapter 4 describes issues beyond the financial crisis, and Chapter 5 concludes with future perspective and policy implications.

Economic Background

Since the economic history of these four countries goes back centuries, it is impossible to cover the entire spectrum in this work. As such, this background discussion will consider events in the relative recent history from 2005 onward that have the most direct impact upon the current economic conditions of each specific country. In addition, particular attention will be given to the 2008 global financial crisis. In doing so, the authors hope that the background, in conjunction with our later data and analysis, will provide a "before, during, and after" picture of what was happening from a macroeconomic and intangible assets viewpoint.

In the year 2011, these four Southern European countries were widely known for their faltering economies and burgeoning national debt. To paint a general picture about their global competitiveness in the most recent years and before the financial crisis, we introduce hereunder the Global Competitiveness Index (GCI) published by the World Economic Forum for readers' reference. This index is relatively robust, for it takes into account the 12 distinct pillars[1] containing basic requirements, efficiency enhancers, and innovation factors that contribute to a nation's overall economic strength. Based upon commonly accepted economic theory, the development of a total 142 countries was split into three stages in which different factors play the dominant role in determining the outcome of a country's economy. Stage 1, Stage 2, and Stage 3 are respectively characterized by being factor, efficiency, and innovation driven (Schwab 2011). Table 1.1 exhibits the GCI ranking of 28 Stage 3 countries out of our studied 48 countries. The table shows that Spain #36, Italy #43, Portugal #45, and Greece #90 are all at the bottom of the list, with Greece being well below the rest.

Table 1.1 Global competitiveness index 2011–2012 ranking of Stage 3 countries (Total 28 countries out of 48 countries)

Stage	Country	GCI ranking	Stage	Country	GCI ranking
3	Switzerland	1	3	Norway	16
3	Singapore	2	3	France	18
3	Sweden	3	3	Austria	19
3	Finland	4	3	Australia	20
3	USA	5	3	Israel	22
3	Germany	6	3	S. Korea	24
3	Netherlands	7	3	New Zealand	25
3	Denmark	8	3	Ireland	29
3	Japan	9	3	Iceland	30
3	United Kingdom	10	3	Spain	36
3	Hong Kong SAR	11	3	Czech Republic	38
3	Canada	12	3	Italy	43
3	Taiwan	13	3	Portugal	45
3	Belgium	15	3	Greece	90

[1] The 12 pillars include institutions, infrastructure, macroeconomic environment, health and primary education, higher education and training, goods market efficiency, labor market efficiency, financial market development, technological readiness, market size, business sophistication, and innovation.

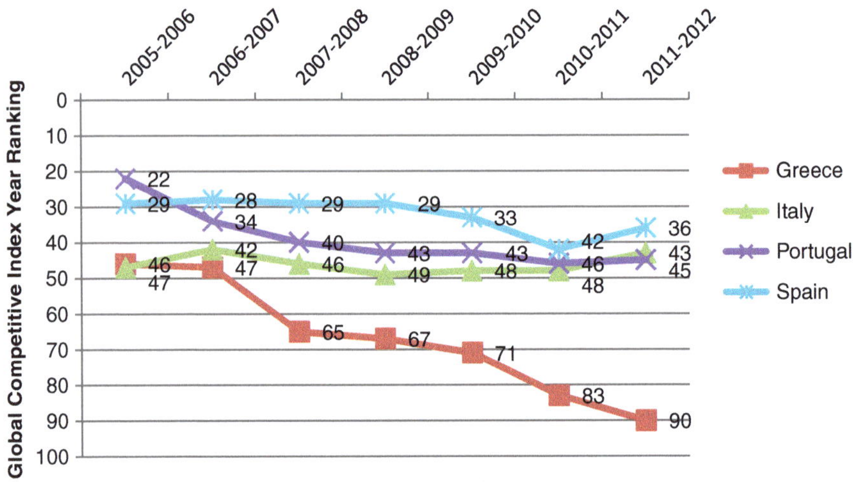

Fig. 1.1 GCI ranking of the four Southern European countries

Taking each country's annual ranking and plotting against a time series of seven periods, Fig. 1.1 displays a rough pictorial overview of the four countries' global competitiveness changes before and after the financial crisis

From Fig. 1.1, it can be seen that over the past seven years, the four Southern European countries have been stagnant or declining with Greece, Portugal, and Spain taking downturns. This is particularly true in the case of Greece. This general trend coincides with the fallen economy phenomenon that these four countries have been experiencing. Specifically, the downward sloping in 2010–2011 (except Italy) represents the effects of the 2010 sovereign debt crisis in Europe.

In early 2012, the economic situation in Southern Europe was bleak. Fiscal austerity imposes a vicious cycle on weaker nations. Budget restraint retards growth, creates still larger deficits that force more restraint, and investors begin to wonder if these nations can ever reach solvency (Ezrati 2012; The Economist 2012). A potential remedy is to pursue a growth strategy in addition to austerity; both strategies can simultaneously exist according to IMF (Ezrati 2012). Measures may include loosening restrictive labor laws, revising growth-stifling tax codes, easing regulatory restrictions, and privatization (Ezrati 2012).

Economic turmoil that surfaced during and after the financial crisis in these four countries has decade-long deep-rooted problems. The next chapter will give a brief background and qualitative analysis of the 2008 global financial crisis as it relates to these four countries as a whole and individually.

Chapter 2
Impact of the 2008 Global Financial Crisis

As mentioned earlier, the 2008 financial crisis rapidly developed and spread into a global economic shock, which resulted in a number of European bank failures and stock market declines (Fackler 2008; Altman 2009). Economies worldwide slowed during this period due to tightening credit and drops in international trade. At the onset of the financial crisis, some European countries viewed it as a purely American phenomenon. Yet, that view quickly changed following the rapid decline of economic activity in Europe. What made matters worse was that global trade shrank sharply, thus eroding the prospects for European exports. In addition, public protests, sparked by rising rates of unemployment and concerns over the growing financial and economic turmoil, increased the political stakes for European governments and their leaders. As a result, the global economic crisis strained the ties that bound together the members of the European Union (EU) and presented a significant challenge to the ideals of solidarity and common interests (Nanto 2009).

Following the outbreak of the financial crisis, on November 26, 2008, the European Union proposed a European stimulus plan amounting to US$256 billion (€200 billion based on 11/26/08 exchange rate) or 1.5% of the EU GDP—around 1.2% of GDP from national budgets and the rest 0.3% of GDP from EU and European Investment Bank budgets (Europa 2008). The stimulus aimed at limiting the economic slowdown through national economic policies extending over a period of two years. The measures include supporting medium-term growth through increased public spending on infrastructures (road networks and railway), assisting the housing sector (notably construction and renovation), and increasing in benefits and allowances to low-income and unemployed households. For the entire world, the estimated US$2 trillion total in stimulus packages amounts to approximately 3% of world Gross Domestic Product, exceeding the call by the International Monetary Fund (IMF) for fiscal stimulus by 2% of global GDP to counter worsening economic conditions worldwide (Nanto 2009).

As of mid-2012, several European countries are still suffering from the sovereign debt problem and need international assistance. In hindsight, statistics revealed that some Euro zone countries are much more heavily indebted than others, as a result

C.Y.-Y. Lin et al., *National Intellectual Capital and the Financial Crisis in Greece, Italy, Portugal, and Spain*, SpringerBriefs in Economics 7,
DOI 10.1007/978-1-4614-5990-3_2, © The Author(s) 2013

of borrowing recklessly at the cheap interest rates available inside the Euro (Cecchetti et al. 2011). The escalating debt threatened to reduce their economies to a fragile level. When the financial crisis hit, they were powerless to fend it off and a sovereign debt crisis swiftly emerged.

The debt problem in these four southern European countries traces its roots back to the previous decade. Loans to the real economy increased at an average rate of around 20% in Greece and Spain over the period 2002 to 2007. During this period, weak financial regulation and supervision encouraged a significant increase in risk taking. The very low real interest rates, which were sometimes even negative, created highly favorable conditions for housing bubbles, as is the case with Spain. This then, increased the current account deficits so that by 2008, Greece, Portugal, and Spain had accumulated net foreign liabilities of over 70% of the national GDP, according to OECD (Barnes 2010). These three countries had a credibility problem. They lacked the ability to adequately repay their debts due to low growth rate, high deficit, and decreasing foreign direct investment (First Post 2011). Spain's creditors were mainly domestic institutions, but Greece and Portugal had a higher percent of their debt in the hands of foreign creditors, which was seen by certain analysts as more difficult to sustain. In addition, Italian national debt had reached around US$2.0 trillion (€1.6 trillion)—three times greater than the national debt of Greece, Portugal, and Ireland combined (First Post 2011).

Furthermore, by joining the Euro zone, member nations gave up the ability to individually devalue their currency in bad times. In the past, countries often used devaluation when there was an economic downturn to balance their economies. To remedy potential problems, the Economic Monetary Union stipulated the Stability and Growth Pact that required member countries to maintain a government budget deficit of no more than 3% of GDP and public debt of no more than 60% of GDP. However, there were no fixed rules governing the exact details of how penalties are levied (Harrington 2011). As a result, most of the countries in deep trouble during this financial crisis had little incentive to follow the rules and were frequently in violation of them.

The most serious Greek sovereign debt crisis was set off in late 2009 when the government admitted that its deficit would be 12.7% of its gross domestic product, not the 3.7% the previous government had forecasted earlier (The New York Times 2012). In early 2010, fears over a potential default grew into a full-fledged financial panic. As the fear spread to Portugal and Spain, leaders of Europe's more affluent countries like Germany and France, worrying about lasting damage to the euro, stepped in with a pledge to defend the common currency. When the problem spread, Greece secured a bailout package worth US$158 billion (€110 billion, based on exchange rate at that time) in May 2010 and then in November 2010 Ireland received a bailout amounting to US$113 billion (€85 billion) (Minder 2011; The New York Times 2012). In April 2011, Portugal was also in trouble and requested an international aid package worth US$110 billion (€78 billion) which was approved in May 2011 (Minder 2011).

As the sovereign debt problems continued to ail the Euro zone, in September 2011, OECD reported that major economies were tilting back into recession and fears were growing that Italy might be the next country facing a financial crisis—after

Greece, Ireland, and Portugal (Bryant 2011). About a year after the initial bailout, the Greek economy continued to sag under US$483 billion (€340 billion) of debt, as the austerity package sent the economy deeper into recession (The New York Times 2012). As a result, a second bailout became necessary. In February 2012, European finance ministers approved a new bailout of US$172 billion (€130 billion euros), subjecting to Greece taking immediate steps to implement the deep structural changes needed to rectify the defunct economy (The New York Times 2012). Spain became the fourth and largest country to ask Europe to rescue its failing banks, with a loan of up to US$125 billion (€100 billion) agreed on by the Euro zone financial ministers on June 10, 2012 (Wolls and Dilorenzo 2012). Since the deal imposed no conditions on the overall Spanish economy and no new austerity measures, the Spanish government emphasized that it was an aid not a rescue (RTE 2012).

The impact of the 2008 global financial crisis on each country can be easily observed from the following four graphs, namely the percentage of real GDP growth per capita, total general government debt percentage of GDP, unemployment rate of labor force, and consumer price inflation from 2005 to 2010.

Comparisons of the Four Countries

Figure 2.1 shows that all four countries had an obvious GDP growth decline starting in 2008 and reached its lowest level in 2009. The exception is Greece, which experiences a continuous drop throughout. In 2010, all but Greece had a clear financial recovery, although Spain still had a negative growth.

In terms of the total general government debt as a percentage of the GDP, Fig. 2.2 indicates that government debt increased from 2008 onwards for all four countries, reflecting increasing financial needs during and after the financial crisis. In particular, the government debt in Italy is relatively stable, as it was already at a relatively high level (over 100%), and Greece saw a sharper debt increase in 2010 compared to the other three countries.

Reinhart and Rogoff (2009) reported findings from their research on financial crises over the last 800 years that the aftermath of a financial crisis brings slow and halted growth, sustained high unemployment, and surging public debt—with the overhang of public and private debt being the most important impediment to a normal recovery from the recession.

As for the unemployment rate, Fig. 2.3 shows that all four countries had unemployment rate increase starting from 2008, with Spain experiencing a sharper increase. A clear sign of warning is that Spain more than doubled its unemployment rate over the six years.

Figure 2.4 indicates the upward trend of consumer price inflation from 2007 to 2008 in these four economies, with Portugal having the least change. However, all inflation dropped drastically in 2009, which very likely came about because of the effect of the financial stimulus plan adopted by each country. Yet, it rose again in 2010 with Greece having the largest increase.

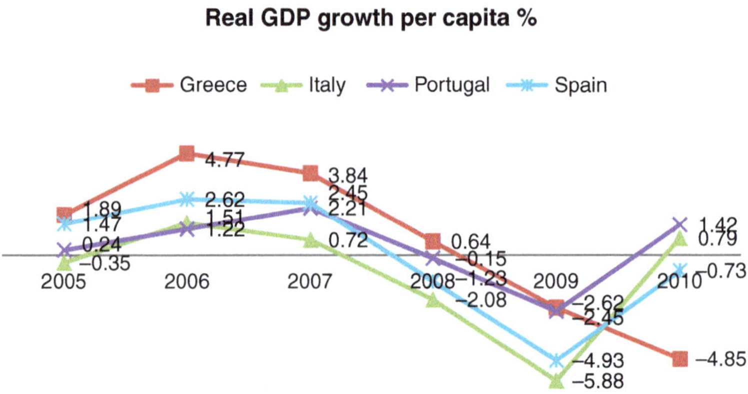

Fig. 2.1 Real GDP growth per capita for Greece, Italy, Portugal, and Spain from 2005–2010

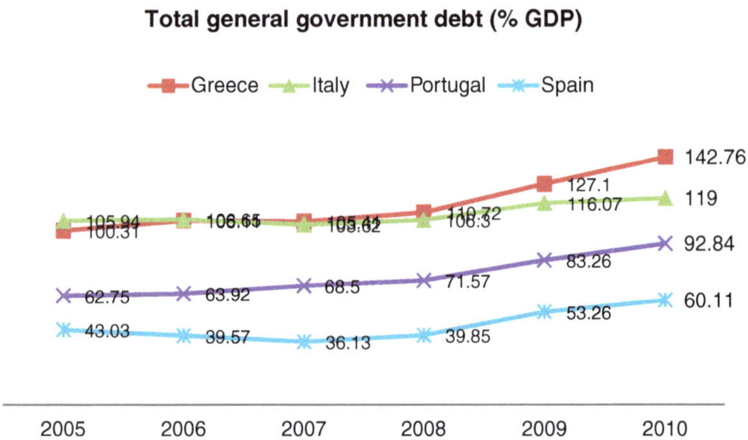

Fig. 2.2 Total general government debt (% of GDP) of Greece, Italy, Portugal, and Spain from 2005–2010

In general, the above four figures for these Southern European countries show that the four indicators were impacted by the 2008 global financial crisis, with the real GDP growth and consumer price inflation having clear fluctuations. The development patterns of these four countries are similar, with Greece having the most negative result and Spain having high unemployment rate yet a low total general government debt.

In what follows, we briefly described the impact of 2008 global financial crisis on Greece, Italy, Portugal, and Spain in sequence. Please note that the reported stimulus package is based on publicly available data and is not an exhaustive list. The reported amount was based on the exchange rate at the time of each stimulus, and thus varies. In addition, the depth of the report depends on the English literature

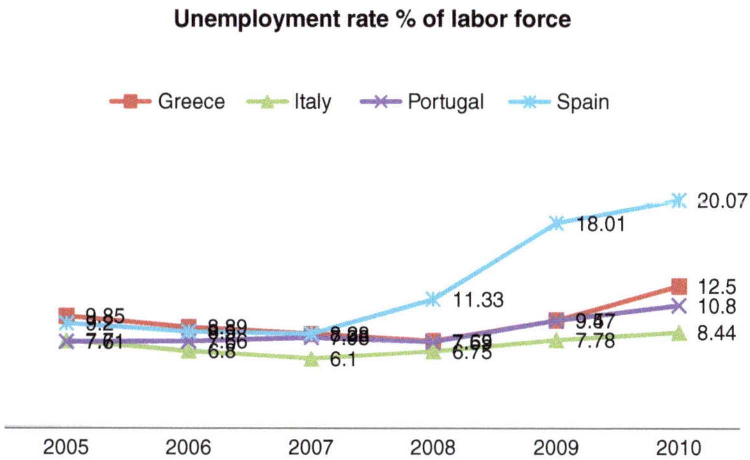

Fig. 2.3 Unemployment rate percentage of labor force in Greece, Italy, Portugal, and Spain from 2005–2010

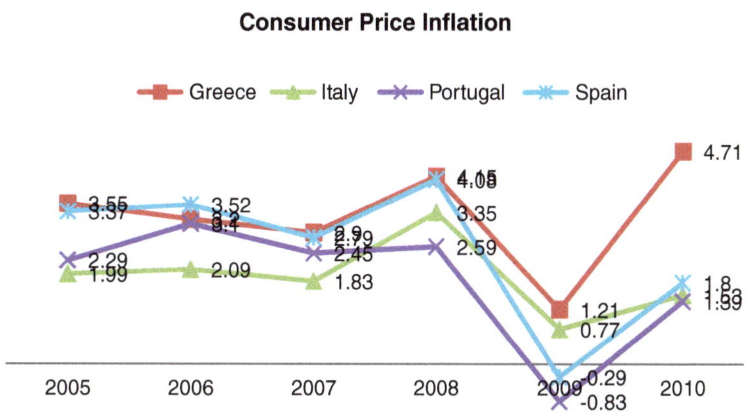

Fig. 2.4 Consumer price inflation of Greece, Italy, Portugal, and Spain from 2005–2010

available for each country. Readers can refer to Appendix 1 for details on the stimulus package each country adopted and Appendix 2 for the important meetings conducted by key global leaders during this financial crisis.

Greece

Greece has a capitalist economy with the public sector accounting for about 40% of GDP (CIA 2012). According to the World Bank, Greece is an advanced country with the service sector making up about 78.8% of its national economic output. It is a major beneficiary of EU aid, equal to about 3.3% of the annual GDP. The Greek economy grew by nearly 4% per year between 2003 and 2007, due partly to infrastructural spending related to the 2004 Athens Olympic Games and in part to an increased availability of credit, supporting record levels of consumer spending (CIA 2012).

During the 1990s, the Greek economy had shown significant progress in terms of fiscal and monetary adjustments, which culminated in its accession to the Economic and Monetary Union (EMU) of the European Union. Post-accession however, the reform zeal subsided and complacency prevailed. Aided by cheap money, the country had not undertaken far-reaching structural reforms since then (Blavoukos and Pagoulatos 2008). Instead, the Greek government borrowed heavily and went on a spending spree after it adopted the Euro. Public spending soared and public sector wages practically doubled in the past decade (Nelson et al. 2011). However, as the money flowed out of the government's coffers, tax income was hit because of widespread tax evasion.

As a result, the fiscal outlook deteriorated significantly in the 2000s. In particular, the 2004 Olympic Games generated deficits, and the government failed to exercise fiscal discipline or implement the necessary reforms in the pension system, public administration, and elsewhere (Pagoulatos 2010). Although from 2000 to 2007, Greece still experienced enviable economic gains with an average real GDP growth of 4.27%, it depended upon being able to successfully finance all of its borrowings to sustain such growth. From 2002 to 2007, Greece violated the EU's Growth and Stability Pact budget deficit criterion of no more than 3% of the GDP with deficits averaging 5.4% of its GDP (Barnes 2010). With the Euro currency, Greece can no longer devalue its own currency in order to finance its debt.

Initially, Greece held up better during the 2008 global economic crisis than many other OECD countries, as the banking sector had only marginal exposure to the toxic assets and growth remained positive until the end of 2008 (Sorsa et al. 2009). Yet, it was unlikely to avoid a recession as global confidence, tourism, and shipping receipts had all fallen substantially. The impact of the crisis shook the confidence of households and businesses, which were reigning in spending. When the financial crisis perpetrated through the entire world, Greece's situation started to become more and more perilous. The economy went into recession in 2009 and contracted by 2% as a result of the world financial crisis and tightening credit conditions (Dimireva 2009). The authorities responded with fiscal measures to assist the financial sector and announced in mid-June 2009 a consolidation plan, aimed at reducing the structural budget deficit by cutting civil service employment, freezing government wages, cutting 10% of "elastic" budget outlays, and levying taxes on high incomes (Sorsa et al. 2009). However, their room for policy maneuver was tightly restricted by the high public debt, repeated fiscal slippages, and large external and internal imbalances, which had been reflected in high sovereign interest-rate spreads as risk aversion rose (Sorsa et al. 2009). For Greece, it was the combination of the high debt to GDP ratio, deficit to GDP ratio, stagnant economy, shrinking tax base, and a dysfunctional tax collection system which increased Greece's vulnerability and exacerbated its shortage of liquidity (Abboushi 2011).

Later on, Greece's eroding public finances, inaccurate and misreported statistics, and consistent underperformance on reforms prompted major credit rating agencies in late 2009 to downgrade its international debt rating. Under intense pressure from the EU and international market participants, the government adopted a medium-term austerity program that includes cutting government spending further, decreasing tax evasion, reworking the health-care and pension systems, and reforming the labor and product markets (CIA 2012). Athens, however, faces long-term challenges to push through unpopular reforms in the face of widespread unrest from the country's powerful labor unions and the general public (CIA 2012). By the end of 2009, as a result of a combination of international and local factors (the world financial crisis and uncontrolled government spending, respectively), the Greek economy faced its most-severe crisis since the restoration of democracy in 1974 (Lynn 2011).

In December 2009, Greece admitted that its debts had reached US$429 billion (€300 billion)—the highest in modern history; in addition, it was burdened with debt amounting to 113% of the GDP—nearly double the Euro zone limit of 60% (BBC 2012a).

Italy

Italy has a diversified industrial economy, divided into a developed and private-company-clustered industrial north and a less-developed and welfare-dependent agricultural south that has high unemployment. Italy has a sizable underground economy, which by estimation accounts for as much as 17% of GDP (CIA 2012). In the mid-1990s, Italy enjoyed vigorous growth and became one of the first 11 Euro zone countries. However, from the late 1990s, it plunged into a downturn, where in most years, economic growth was behind that of Europe's average. In 2005, it had the worst economic statistics of all Euro zone countries: zero growth in GDP and the deficit jumped to 4.1% of the GDP; in 2006, because of the Torino Olympic Winter Games, there was a rise of 1.9% in GDP; in 2007, Italy's GDP rose 1.5%, while the European Union enjoyed a 2.9% increase (Wang 2009).

In addition, Italy has had a debt ratio of over 100% of its GDP ever since 1991. The country averaged an abysmal 0.75% annual economic growth rate over the past 15 years, which was much lower than the rate of interest it pays on its debts (Knight 2011). The main cause of Italy's "current account deficit" was the weak Italian economic performance resulting from structural problems (Barnes 2010). Like other southern European economies, Italian wage levels rose too quickly during the good years, and left Italy uncompetitive within the Euro zone.

Italy, unlike Greece, actually had been quite financially prudent because the government spent less on providing public services and benefits to its people than it earned in taxes (Knight 2011). The only reason Italy continued to borrow was to meet the principal and interest payments on its existing debts. Furthermore, it was plagued by poor regulation, vested business interests, an aging population, and weak investments. All of which conspired to limit the country's ability to increase production (Knight 2011). As a result, the high public debt burdens and structural impediments to growth rendered Italy vulnerable to scrutiny by financial markets (CIA 2012).

When the financial crisis exploded in autumn 2008, the Italian Government was prudent. The first stimulus package amounting to US$3.8 billion (€3 billion) was introduced in November 2008. It included transfers to low-income households and relief measures for enterprises, fully financed by revenue increases and expenditure cuts (Dilje 2008; Economy Watch 2010a). Following the European Economic Recovery Plan, a second fiscal package was approved in February 2009. The total value of this economic stimulus package was worth around US$101 billion (€80 billion) (Economy Watch 2010a; OECD n.d.) The package included a temporary freeze on regulated energy prices and road tolls, around US$3 billion (€2.4 billion) in tax breaks for poorer families, reduction in advance tax payments and speedy reimbursements of excess tax payments, some marginal easing of the direct and indirect tax burden for companies, delayed payment of VAT, a car-scrapping incentive, and mortgage rates capped at 4% (Dilje 2008; OECD n.d.). This package also emphasized the state's responsibility of underwriting special convertible bonds issued by banks, including state help for Italian banks worth up to US$25.2 billion (€20 billion) to ensure their continuous support by lending (Economy Watch 2010a).

In general, 2009 was characterized by widespread negative economic performance. GDP in Italy decreased in real terms by five percentage points from its average value recorded in 2008 (Knight 2011). In 2009, the number of people in employment declined by 380,000 (−1.6% on an annual basis), while the unemployment rate rose to 7.8% (Eurofound 2010). The economic recession affected the Italian production system in its entirety, with mechanical engineering and small companies taking the worst hit in terms of turnover, investments, and employment. In many cases, these crises also affected large-sized companies, causing damaging effects to subcontractors and suppliers (Eurofound 2010). During the financial crisis, the largest job losses have taken place among workers on temporary employment contracts.

Overall, Italy's precrisis stagnation and crisis vulnerability resulted from an inefficient state of bureaucracy, high levels of corruption, heavy taxation, and high public spending that accounts for about half of the national GDP. Like Spain, the country's productivity level was low, and like Greece, it had a problem in collecting tax and suffered from massive debt (First Post, n.d.).

Portugal

Portugal has become a diversified and increasingly service-based economy since joining the European Community—the EU's predecessor—in 1986; after that, the economy grew by more than the EU average for much of the 1990s, but fell back in 2001–2008 (CIA 2012). Since then, Portugal has become one of the weakest economies in the Euro zone, with traditionally high unemployment, inflexible labor laws, and governments that paid scant attention to improving overall competitiveness (Reguly 2011). From 2002 to 2007, deficits averaged well over 3% of GDP and the unemployment rate increased 65% (270,500 unemployed citizens in 2002, 448,600 unemployed citizens in 2007) (Barnes 2010). As a result, the country was ill prepared to get through the financial crisis of 2008.

Initially, the financial sector actually remained sound in Portugal, with the absence of a real estate bubble in the years preceding the crisis and low exposure to toxic assets (Barnes 2010). However, the country has been increasingly overshadowed by lower-cost producers in Central Europe and Asia as a target for foreign direct investment (CIA 2011). In 2008, Portuguese economic growth remained positive at 0.3%. Yet, the cumulating government debt, lasting high unemployment, high public spending, and low productivity began to crack up with the unfolding of the global financial recession. The GDP of Portugal contracted by 2.7% in 2008–2009 largely due to shrinking domestic demand. The impact on the labor market and on public finances has been especially severe, as unemployment rose to 9.47% in 2009; the budget deficit increased from 2.8% of GDP in 2008 to 9.4% in 2009 (Euro Challenge 2012).

This weak position is partly the result of overreliance on consumption and housing activity in the early 2000s. Weak subsequent labor productivity coupled with insufficient wage moderation has caused the large current account deficits (Barnes 2010). Portugal's growth prospect was weak and the financial crisis affected its economy severely, causing a wide range of domestic problems. Public deficit hiked with excessive debt levels, soaring to about 223% of Portugal's GDP (Tirone 2011). While the sovereign debt crisis took place mainly in Greece, it also triggered a wider lowering of confidence in national government throughout the entire European Union. As Portugal is akin to Greece in high government deficits and runs a deficit of 9.4% of GDP, it was impacted severely.

The Portuguese response to the crisis, approved by the Council of Ministers in December 2008, was the Investment and Employment Initiative. This strategy included an enhanced set of measures focused on the reinforcement of public investment, modernization of schools, technological infrastructures, direct support to the economy through fiscal measures, incentives for SMEs and exports, and support for employment and for the adoption of renewable energies (Economy Watch 2010b; Nelson et al. 2011). The Investment and Employment Initiative represents an additional stimulus to the economy of around US$2.92 billion (€2.18 billion, about 1.25% of GDP), of which US$1.74 billion (€1.3 billion, 0.8% of GDP) is directly financed through the government budget. Fifty percent of Portugal's economic stimulus package would be provided by members of European Union and remaining amount would come from Portugal itself (European Commission 2010; Economy Watch 2010b).

In January 2009, the Portuguese Parliament approved a new scheme of fiscal incentives focused particularly on research and development (R&D), extending the maximum rate of tax credit to 82.5% of total expenses on R&D, the highest rate in Europe. The system comprises two distinct components, cumulative in nature, with a fixed tax credit of 32.5% of total yearly expenses on R&D (also the highest in Europe), together with a second component of 50% over the annual increase of those expenses (OECD 2009). Unfortunately, the crisis coping measures did not turn the economy around.

In December 2009, Standard and Poor's rating agency lowered its long-term credit assessment of Portugal to "negative" from "stable," voicing concerns about the country's structural weaknesses in its economy and decreasing competitiveness (First Post 2011). Investors bet against Portugal, raising their premiums, and making it increasingly likely the country would not be able to finance itself in debt markets. Overall, the country's dependence on foreign debt made it more susceptible to the financial crisis.

Spain

Spain's mixed capitalistic economy is the 13th largest in the world, and its per capita income roughly matches that of Germany and France; however, after almost 15 years of above average GDP growth, the Spanish economy began to slow in late 2007 and entered into a recession in the second quarter of 2008 (CIA 2012). Of all the European countries, Spain was one of the most gravely affected by the global economic crisis. Even without the global recession, Spain would most likely be undergoing a correction due to its extremely overheated housing market. At one point, investments in the Spanish housing sector made up almost 10% of its GDP (Stratfor 2009). Spanish mortgage lenders were offering loans very liberally—particularly to young migrants with no prior credit history, as part of government policies to speed up integration and assimilation—and were often giving loans of more than 100% of a property's total value (Stratfor 2009). In addition to heavily subsidized home ownership, wage growth outpaced that of other European countries, thus heightening financial vulnerability. During the third quarter of 2008 in Spain, the national GDP contracted for the first time in 15 years and in February 2009, it was confirmed that Spain had officially entered into a recession.

Actually, the Spanish banking system has been credited as one of the most solid and best equipped among all Western economies to cope with the worldwide liquidity crisis, as banks are required to have high capital provisions. In addition, banks demand various proofs and securities from intending borrowers (The Economist 2008). However, according to the Spanish Ministry of Housing, the residential real estate prices rose 201% from 1985 to 2007 and there was a growing family indebtedness (115%) chiefly related to the real estate boom and rocketing oil prices. Government housing statistics also showed that in the second quarter of 2005, Spanish families already carried US$841,700 million (€651,168 million) in mortgage debt. Based on Eurostat, from June 2007 to June 2008 Spain was the European country with the sharpest plunge in construction rates. After the outbreak of the 2008 global financial crisis, Spain experienced financial worries as well due to the collapse of its housing market, the deep involvement of unregulated savings banks in the real estate market, and increased public benefits that caused higher government spending and government debt.

Another key factor that complicated the situation was that around half of Spanish banks were unregulated. Spanish regional savings and loan banks, called *cajas*, account for half of Spain's banking system. They are not publically traded and are usually controlled by regional politicians instead of shareholders; in addition, they were not required to disclose certain information such as collateral on loans, repayment history, and loan-to-value ratios (Harrington 2011). Cajas often loan to those that larger banks turn away due to worries of repayment ability. However, when Spain's two largest banks slowed lending in 2007, the cajas continued to lend heavily into the cooling housing market. By 2009, cajas owned 56% of the country's mortgages, and loan payments from property developers accounted for one-fifth of the cajas' assets (Harrington 2011). When the housing market crashed in 2009, the cajas were paralyzed by a lack of income from these delinquent loans (Harrington 2011).

According to Hugh (2010), Spain's basic problem is not a fiscal one but a rapid escalation of the fiscal deficit in the context of a heavily indebted private sector. Even before the financial crisis, Spain's unemployment rate climbed around 37% between October 2007 and October 2008. By July 2009, it had shed 1.2 million jobs in one year. Compared to February 2008, Spain's industrial production declined 22% in February 2009 (Stratfor 2009). According to National Statistics Institute released on April 24, 2009, Spain for the first time in her history had over 4,000,000 people unemployed (BBC 2009). As unemployment skyrocketed, so did unemployment benefits. In a welfare state like Spain, unemployment benefits are generous. However, the reductions in the Spanish government's tax revenue, which is heavily dependent on real estate, exacerbated the problem.

In late 2008, the Spanish government announced a stimulus package of US$66 billion (€50 billion). Of that amount, US$10.6 billion (€8 billion) was distributed directly to local authorities for public projects meant to create 200,000 jobs (European Commission 2010; Stratfor 2009). With global demand for auto exports and industrial goods declining, production in the Spanish automotive sector declined by a whopping 47.6% in February 2009. To save the industry, the government gave around US$2,577 (€2000) in subsidies to new car buyers during the financial crisis to support car sales, as Spain's automotive sector accounts for about 10% of the country's total economic output and 15% of its total exports (Stratfor 2009).

In the past decade, Spain has enjoyed the low EU interest rate which prompted its continued growth. Yet by adopting the Euro, Spain could not devalue its currency. With Spain's rising debt, investors became more reluctant to invest and Spain's economy further declined (Harrington 2011).

Chapter 3
National Intellectual Capital Development of the Four Southern European Countries

National Intellectual Capital Development

National intellectual capital measurement model in this chapter is based on the indicators described in Appendix 3, containing human capital, market capital, process capital, renewal capital, and financial capital. This model has been statistically validated for its validity and reliability, and the definitions of the indicators are listed in Appendix 4. This booklet series covers a total of 48 countries in 11-country clusters (see Appendix 5 for details).

For a full picture of all the 48 countries being studied, Appendix 6 exhibits the scores and rankings of human capital, market capital, process capital, renewal capital, financial capital, and overall national intellectual capital (hereafter termed NIC) covered in this booklet series. Table 3.1 summarizes the status of the four Southern European countries compared to the total 48 countries reported in Appendix 6. Appendix 7 provides the individual country profiles with additional statistics. For the NIC ranking, Portugal ranks 26, followed by Spain at 27, Italy at 28, and Greece at 31.

The following six figures compare human capital, market capital, process capital, renewal capital, financial capital, and NIC of the four countries. All individual capital scores are based on a 1–10 scale with the NIC score being the sum of the five capital scores

Human Capital

National human capital mainly represents a nation's investment in nurturing qualified human resources, such as higher education enrollment, skilled labor, and public expenditure on education. It is a long-term oriented NIC (Lin and Edvinsson, 2011: 323), taking time to develop yet not susceptible to external impact. Figure 3.1

C.Y.-Y. Lin et al., *National Intellectual Capital and the Financial Crisis in Greece, Italy, Portugal, and Spain*, SpringerBriefs in Economics 7, DOI 10.1007/978-1-4614-5990-3_3, © The Author(s) 2013

Table 3.1 National intellectual capital scores and ranking of Greece, Italy, Portugal, and Spain among 48 countries spanning 2005–2010

	Human capital		Market capital		Process capital		Renewal capital		Financial capital		NIC	
Country	Score	Ranking	Score	Ranking	Score	Ranking	Score	Ranking	Score	Ranking	Score	Ranking
Greece	5.986	31	4.216	43	4.113	35	2.168	32	9.440	23	25.923	31
Italy	6.471	28	4.155	44	5.184	27	2.745	26	9.479	22	28.035	28
Portugal	6.882	22	4.801	35	5.470	24	2.467	27	9.224	28	28.844	26
Spain	6.524	27	4.649	37	5.033	30	2.787	25	9.484	21	28.476	27

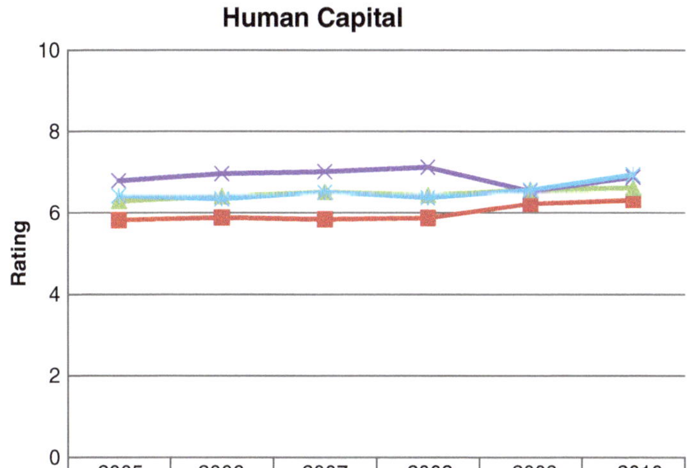

	2005	2006	2007	2008	2009	2010
Greece	5.82	5.88	5.83	5.87	6.22	6.31
Italy	6.29	6.40	6.52	6.44	6.57	6.62
Portugal	6.79	6.96	7.01	7.12	6.53	6.88
Spain	6.42	6.34	6.51	6.37	6.57	6.94

Fig. 3.1 Human capital of Greece, Italy, Portugal, and Spain

shows that human capital of these four countries is relatively stable over the studied 6 years, with all four scores of the last 2 years getting closer, especially those of Portugal and Spain. All countries, except Portugal, increased their human capital after the financial crisis (2009 and 2010). Over the 6 years, Greece had the lowest human capital among the four.

Market Capital

National market capital mainly represents international trade and relationships, such as cross-border ventures, globalization, and exports of goods. Figure 3.2 shows that all the countries had market capital decrease in 2008, with Greece having a larger scale drop and Portugal having a delayed decrease in 2009. In 2010, market capital of these countries bounced back a little, except Greece. Two clusters were formed in 2010, Portugal and Spain as the upper cluster and Greece and Italy the lower cluster.

Process Capital

National process capital represents the infrastructure required for building a prosperous society, such as capital availability, mobile phone subscribers, and government efficiency. Figure 3.3 shows a larger variation and decrease in 2008 with

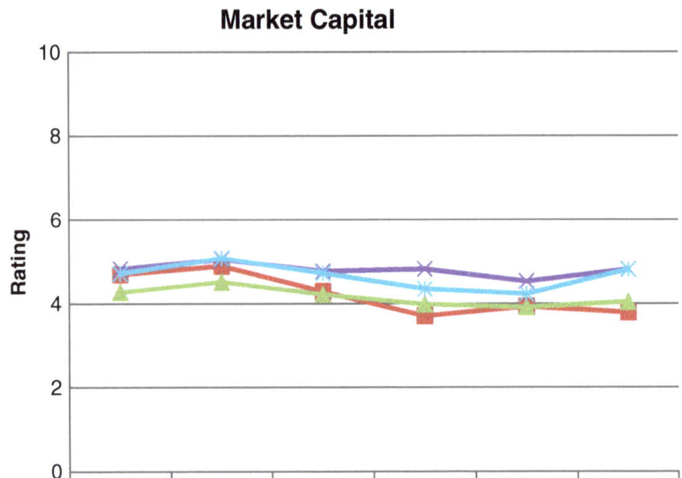

	2005	2006	2007	2008	2009	2010
Greece	4.69	4.89	4.29	3.70	3.93	3.79
Italy	4.27	4.51	4.22	3.98	3.91	4.04
Portugal	4.82	5.05	4.77	4.82	4.53	4.81
Spain	4.71	5.08	4.72	4.35	4.22	4.81

Fig. 3.2 Market capital of Greece, Italy, Portugal, and Spain

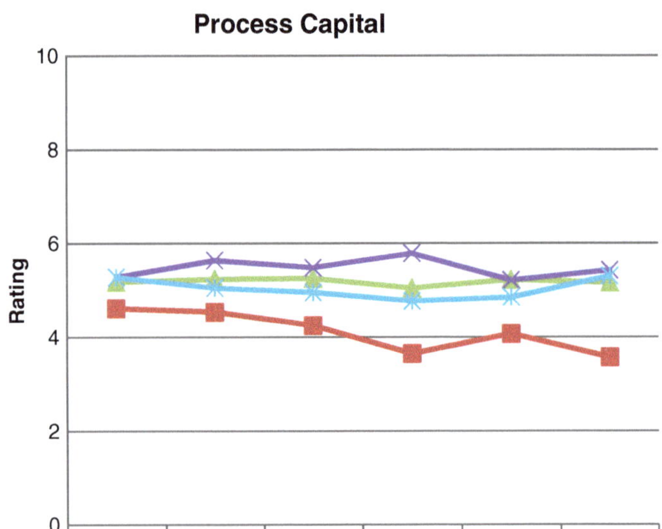

	2005	2006	2007	2008	2009	2010
Greece	4.61	4.54	4.25	3.64	4.07	3.57
Italy	5.17	5.24	5.25	5.05	5.23	5.17
Portugal	5.28	5.64	5.48	5.79	5.21	5.43
Spain	5.27	5.05	4.95	4.77	4.85	5.30

Fig. 3.3 Process capital of Greece, Italy, Portugal, and Spain

Portugal having a delayed decrease in 2009. Greece has the lowest process capital score in this country cluster all through the 6 years, having the largest slide in 2008 and a further decrease in 2010. After the 2008 financial crisis, process capital of Greece slides apart from the other three countries' process capital. This is similar to its separation from the other three countries in terms of Global Competitiveness Index, ranked 36, 43, and 45 for Spain, Italy, and Portugal, respectively, in 2010; yet 90 for Greece as exhibited in Table 1.1 and Fig. 1.1.

Renewal Capital

National renewal capital represents the capability of R&D and innovation, such as R&D spending, scientific articles, and patents. Figure 3.4 shows that except for Greece, starting in 2008, renewal capital of the other three countries either increased or maintained the same level. Portugal particularly had continuous renewal capital increases since 2007 without being affected by the financial crisis. On the contrary, Greece's renewal capital kept declining, separating further away from the rest of the countries in 2010.

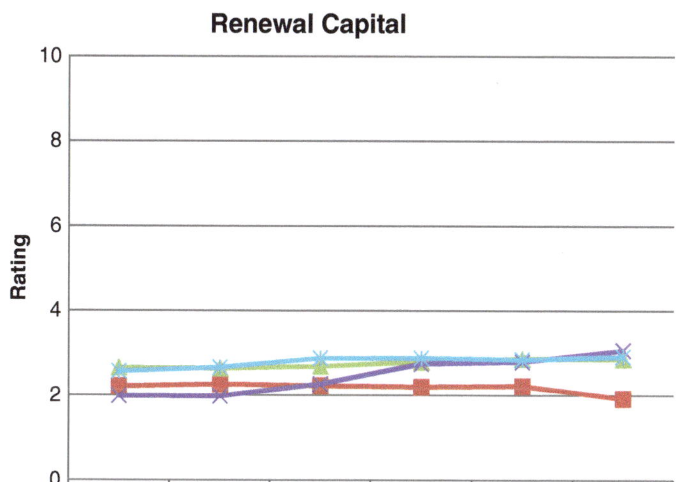

	2005	2006	2007	2008	2009	2010
Greece	2.21	2.26	2.22	2.19	2.22	1.92
Italy	2.65	2.63	2.68	2.79	2.86	2.86
Portugal	1.98	1.97	2.26	2.74	2.79	3.06
Spain	2.57	2.65	2.87	2.87	2.84	2.91

Fig. 3.4 Renewal capital of Greece, Italy, Portugal, and Spain

Financial Capital

National financial capital is the GDP per capita (ppp) in US dollars transformed to a score between 1 and 10. The score transformation process is explained in Appendix 3. Figure 3.5 shows that Portugal lagged far behind the other three countries in financial capital. Yet in real dollars, the average GDP per capita (ppp) from 2005 to 2010 was US$28,223 for Greece, US$29,390 for Italy, US$22,306 for Portugal, and US$29,546 for Spain.

The above analyses reveal that although Portugal ranked the highest in NIC in this country cluster, its financial capital was the lowest. Such information indicates that Portugal did not capitalize on its higher intangible assets to generate more tangible values. In other words, the financial management of Portugal has much room for improvement, which corresponded to its need for EU bailouts during the financial crisis.

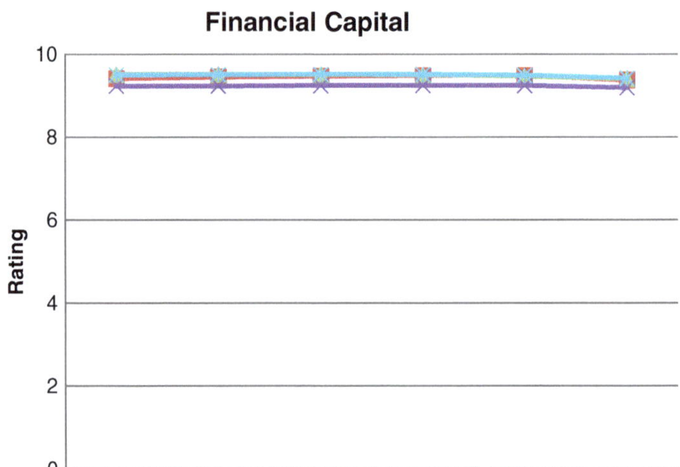

	2005	2006	2007	2008	2009	2010
Greece	9.41	9.44	9.46	9.48	9.48	9.37
Italy	9.51	9.51	9.50	9.49	9.46	9.40
Portugal	9.23	9.23	9.23	9.24	9.24	9.19
Spain	9.49	9.50	9.51	9.50	9.48	9.41

Fig. 3.5 Financial capital of Greece, Italy, Portugal, and Spain

NIC

NIC is the overall national intellectual capital containing the five types of capital discussed above. Figure 3.6 presents NIC trends. Particularly in 2008, there was a larger variation with Greece decreasing drastically and Portugal showing an increase, yet with a delayed decrease in 2009. For Greece, although its NIC increased in 2009, it slid down again in 2010. For Portugal and Spain, NIC increases in 2010 indicate a recovery of their intangible assets.

The Relationship Between Each Individual Capital and GDP Per Capita (ppp)

We investigate the relationship between each individual capital and economic growth by using the following six figures (Figs. 3.7, 3.8, 3.9, 3.10, 3.11, and 3.12). The graphs have four dimensions: the X-axis changes from graph to graph (NIC, human capital, market capital, process capital, and renewal capital), the Y axis is a fixed dimension of GDP per capita (ppp) in US dollars (hereafter termed GDP), the color shade represents renewal capital with red denoting higher levels of renewal

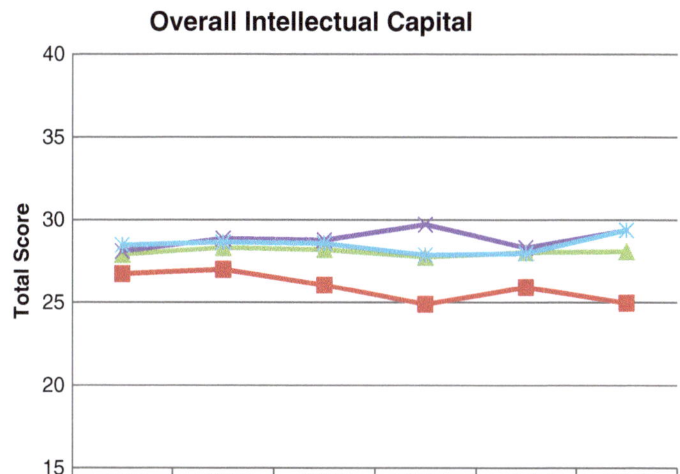

	2005	2006	2007	2008	2009	2010
Greece	26.74	27.00	26.04	24.88	25.91	24.96
Italy	27.89	28.30	28.17	27.74	28.02	28.08
Portugal	28.10	28.85	28.75	29.70	28.30	29.36
Spain	28.47	28.63	28.56	27.87	27.96	29.37

Fig. 3.6 NIC of Greece, Italy, Portugal, and Spain

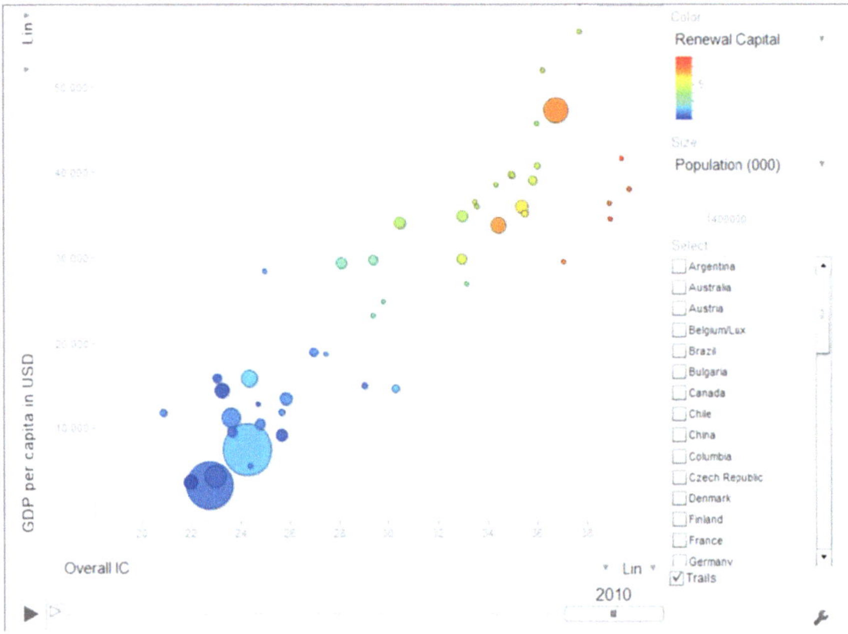

Fig. 3.7 NIC versus GDP Per Capita (ppp) for 48 countries in 2010

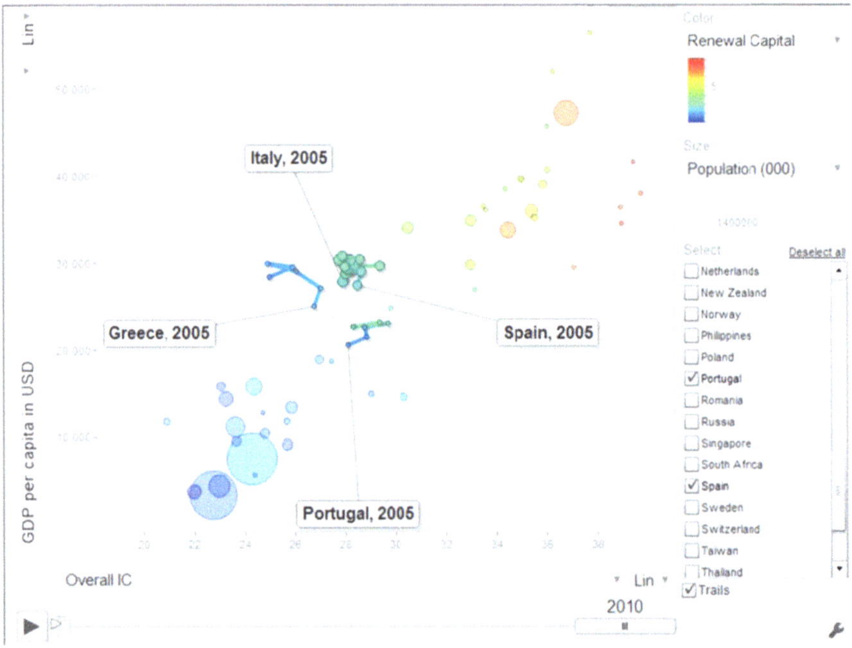

Fig. 3.8 The development of NIC and GDP per capita (ppp) for the four Southern European countries from 2005 to 2010

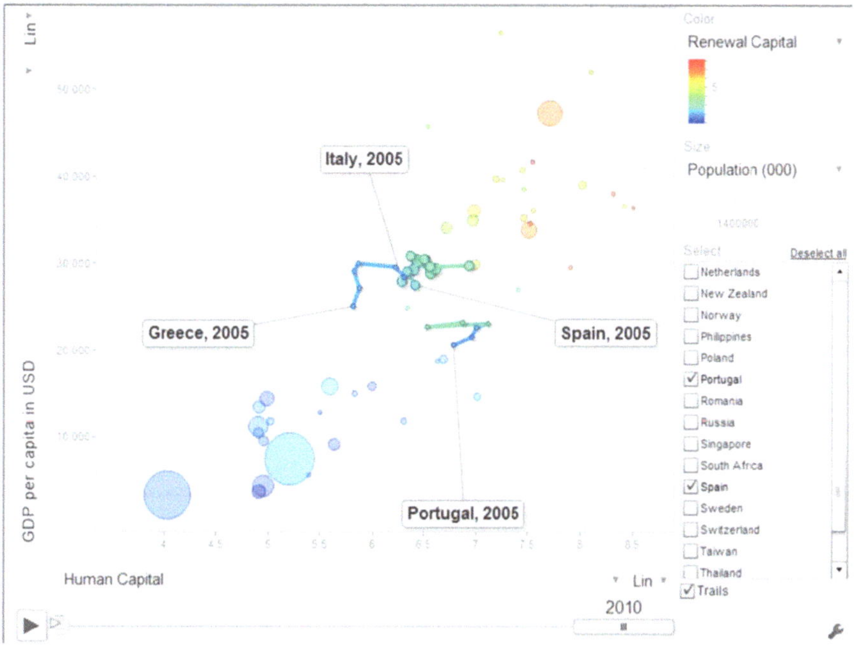

Fig. 3.9 The development of human capital and GDP per capita (ppp) for the four Southern European countries from 2005 to 2010

capital, and the bubble size denoting the population of a nation. Figure 3.7 is the relationship between NIC and GDP for all 48 countries. Figures 3.8, 3.9, 3.10, 3.11, and 3.12 display the codevelopment of the intellectual capital-GDP relationship over the 6 years time frame for the four countries. Since financial capital is the transformed 1–10 score from the GDP, there is no need to plot its relationship with GDP in US dollars. In the figures, year 2005 is the starting point tagged with the name of the countries, and the ending point is year 2010.

Figure 3.7 is particularly plotted to clearly show a full picture of the relationship between NIC and GDP for 48 countries in year 2010. The strong positive correlation indicates that the higher the NIC, the higher the GDP and vice versa. Consequently, enhancing NIC is definitely a strategic issue that deserves the attention of national policy makers.

With the status of all 48 countries as the background, Fig. 3.8 indicates that NIC and GDP codevelopment of the four countries falls within the middle range of the 48 countries. For Italy and Spain, their developments are intertwined. They had similar levels of NIC development which correlated to GDP growth. Greece had the lowest NIC in this country cluster, and its development went backward over the 6 years period. Portugal had higher NIC than Greece with a trend of continuous improvement over the 6 years. Yet, its GDP growth (vertical progression) is less

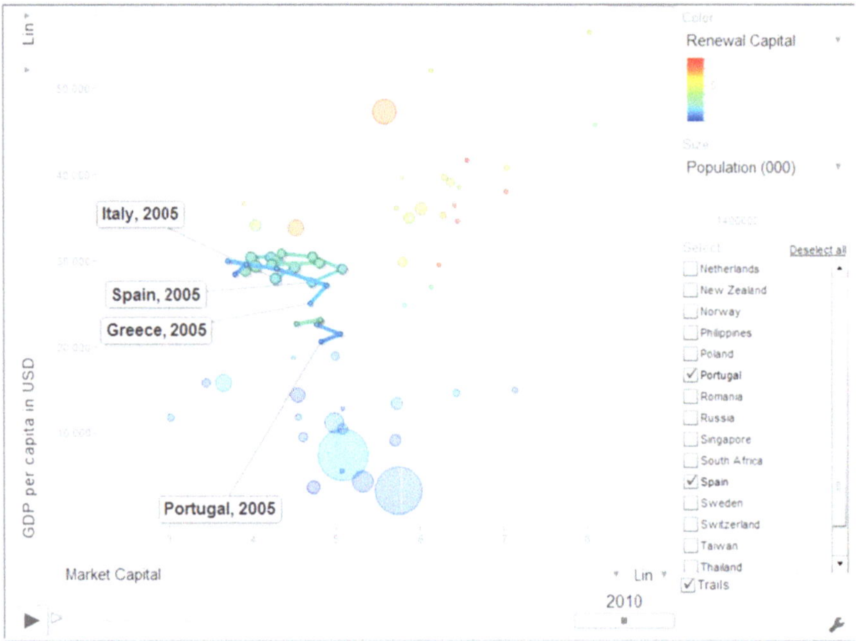

Fig. 3.10 The development of market capital and GDP per capita (ppp) for the four Southern European countries from 2005 to 2010

than that of Greece's. This phenomenon has been highlighted in Fig. 3.5 and poses an interesting issue for further exploration.

Figure 3.9 shows that the codevelopment of human capital and GDP of the four countries is also at the middle range among the 48 countries. The development pattern is somewhat similar to that shown in Fig. 3.8 for NIC, with the exception that Greece had continuous human capital progression, whereas Portugal experienced some human capital regression over the 6 years. For all the countries, the increase of human capital was accompanied with an increase of GDP (vertical progression). For Italy and Spain, the development pattern is similar to that of Fig. 3.8—an intertwined development.

In general, the bubble formation of Fig. 3.10 is very different from that of other figures in this section, without a clear belt shape continuum from the bottom left to the upper right. It means that market capital is not as highly related with GDP as the other capitals are in this set of data. A further explanation is that market capital of advanced countries (generally positioned at the upper right) is not necessarily better than that of emerging countries (the two big bubbles at the bottom center are China and India).

Development of the four Southern European countries is located in the middle left-hand side of the graph. This figure also indicates that all the four countries experienced declining market capital over the 6 years period, with Greece declining the

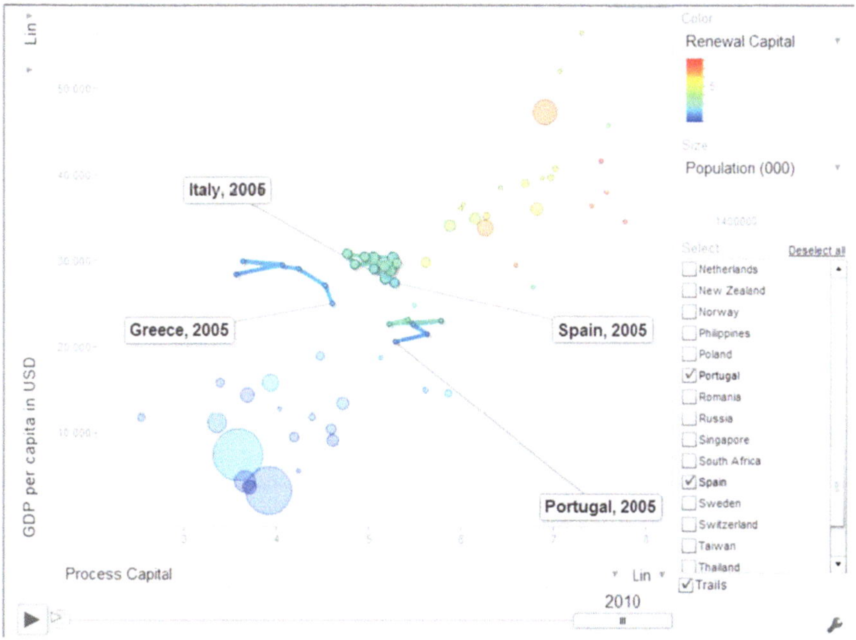

Fig. 3.11 The development of process capital and GDP per capita (ppp) for the four Southern European countries from 2005 to 2010

most, followed by Spain and then Italy. Yet, the declining market capital was still accompanied with GDP growth for all the four countries. A likely answer is that there are multiple sources of GDP growth which pushed up the GDP even though market capital was not an enhancer.

In Fig. 3.11, the belt shape continuum appears again, indicating that process capital is highly related with GDP in general. The development of the four countries is at the middle range among the 48 countries. For Italy and Spain, their developments were again intertwined, with similar levels of process capital correlating to similar levels of GDP growth. Greece had the lowest process capital in this country cluster, and its development backtracked over the 6 years. Portugal had higher process capital than Greece with some improvement over the 6 years, yet its GDP growth (vertical progression) is less than that of Greece's.

Figure 3.12 indicates that renewal capital and GDP codevelopment of the 48 countries had two formations. The spreading formation for the advanced countries in the upper right, and the belt shape continuum for the less-developed countries in the bottom left. That means, for advanced countries, higher renewal capital does not necessarily correspond to higher GDP. Yet, for less-developed countries, increases in renewal capital correlate closely with GDP growth. From the formation of the bottom left continuum, it is clear that small increases in renewal capital were accompanied

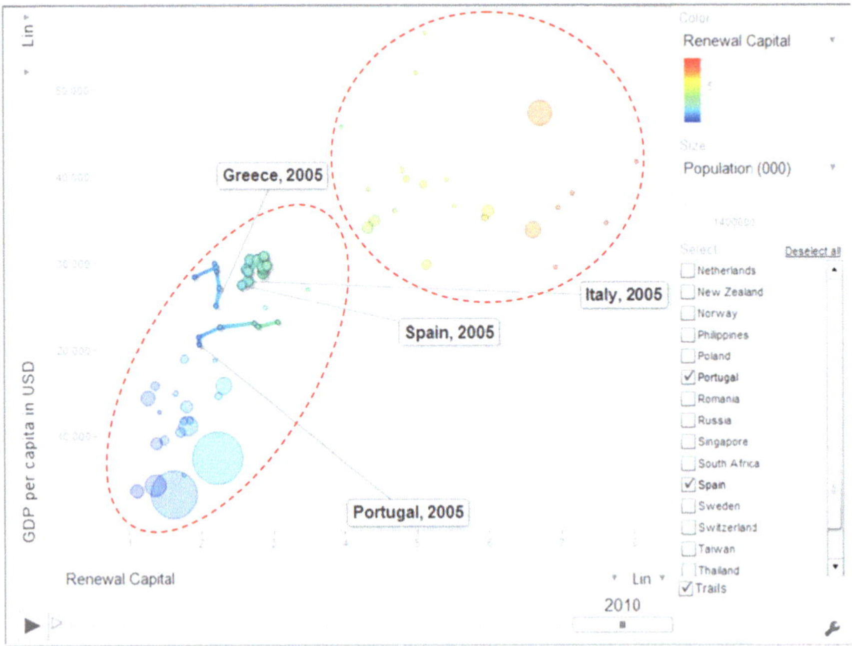

Fig. 3.12 The development of renewal capital and GDP per capita (ppp) for the four Southern European countries from 2005 to 2010

by relatively high GDP growth. This finding reveals a strong message that investing in renewal capital development may benefit GDP growth for less-developed countries.

The renewal capital and GDP codevelopment of the four countries is at the middle to lower range of the 48 countries. Again, the developments of Italy and Spain were intertwined. Greece had better renewal capital than Portugal in 2005, yet it decreased from 2009 to 2010. On the contrary, Portugal started out with lower renewal capital than Greece, yet it improved over the years to higher renewal capital levels than Greece in 2010. However, despite Portugal's higher renewal capital, GDP growth still lagged that of Greece, an interesting issue waiting further exploration.

Although these four countries are all Stage 3 innovation-driven countries according to the World Economic Forum classification (Table 1.1), their renewal capital was grouped together with less-developed countries in Fig. 3.12. This graph uncovers a clear and important direction for future national development of these four countries.

Long-Term and Short-Term National Intellectual Capital

In addition to the above six figures that show the relationship of each individual capital with GDP, Figs. 3.13–3.16 further plot the relationships between human capital and renewal capital, as well as between market capital and process capital. In our first study (Lin and Edvinsson, 2011: 324), we found that the combined increase in human capital and renewal capital would lead to a nation's long-term development, whereas the combined increase in market capital and process capital might result in more immediate economic development. To present their relationships, Figs. 3.13–3.16 map the four capitals on a coordinate with two types of presentation. Based on the same data, Figs. 3.13 and 3.15 exhibit the scatterplot for the four countries only, whereas Figs. 3.14 and 3.16 show all the 48 countries as background so that readers will know the relative positions of these four countries. Please note that the connation of color shade for Figs. 3.14 and 3.16 has been changed from renewal capital to GDP per capita (ppp).

Both Figs. 3.13 and 3.14 indicate that Italy and Spain are at the similar level and intertwined with each other in their long-term NIC development. Greece lags the other three countries in these two long-term oriented capitals. That is, Greece still needs time to catch up with its Southern European peers, especially with regard to renewal capital. Initially, Portugal had higher human capital yet lower renewal capital than the other three countries in 2005. After the financial crisis, its long-term NIC was clustered together with that of Italy and Spain. Hidden information of this development is that high human capital may facilitate the progression of Portugal's renewal capital, as qualified human resources are required for advancing R&D and innovation—major elements of renewal capital.

Figure 3.14 indicates that these four countries are at the middle lower part of long-term NIC development among the 48 countries. Portugal's progress in renewal capital shows promising prospect of revitalizing its troubled economy through enhancing its renewal and innovation power from our findings presented in Fig. 3.12 above. For Greece, unfortunately its improving human capital not only did not generate improving renewal capital but instead saw a backward movement. Such phenomenon may be associated with its economic problem and is a serious warning for coping strategies.

Since these four countries are still in the midst of financial trouble, in addition to the short-term stimulus measures, long-term development of human capital and renewal capital should not be neglected as they are the fundamentals of a nation's sustainability. As human capital is based on unpredictable human performance and renewal capital comes from the outcome of qualified human resources, they take time to develop and should always be on a nation's development agenda.

Figures 3.15 and 3.16 map market capital and process capital of the four countries on a coordinate. Market capital refers to the general assets embodied in a nation's relationship with international markets, while process capital refers to the national infrastructure. For short-term competition, these are the two types of capital that can be expected to produce faster results if a nation is willing to engage in more extensive investments.

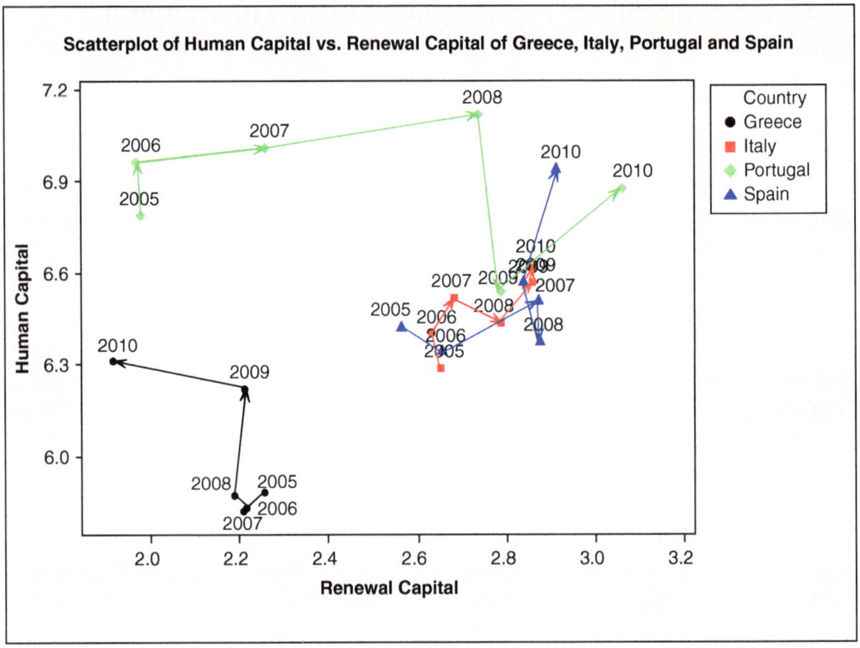

Fig. 3.13 Scatterplot of human capital versus renewal capital for the four Southern European countries

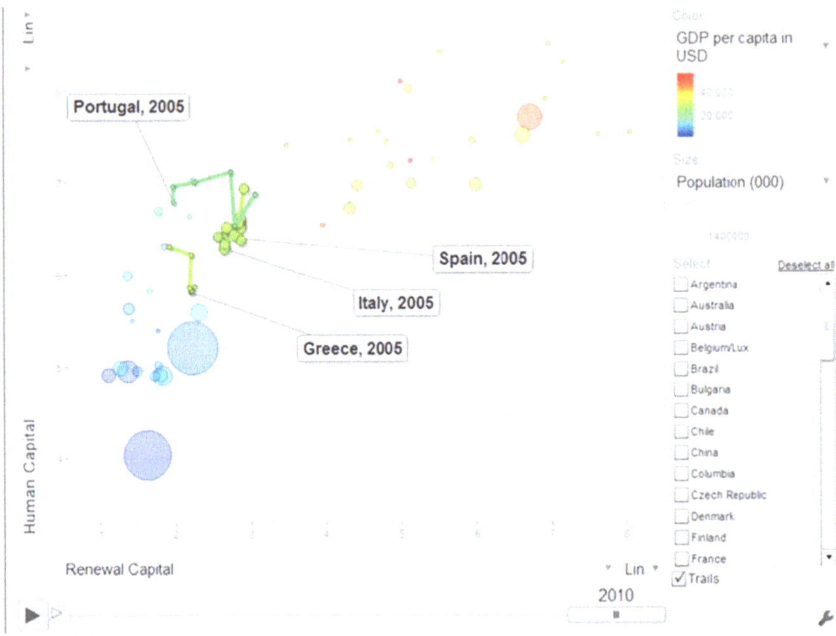

Fig. 3.14 Human capital versus renewal capital for the four Southern European countries

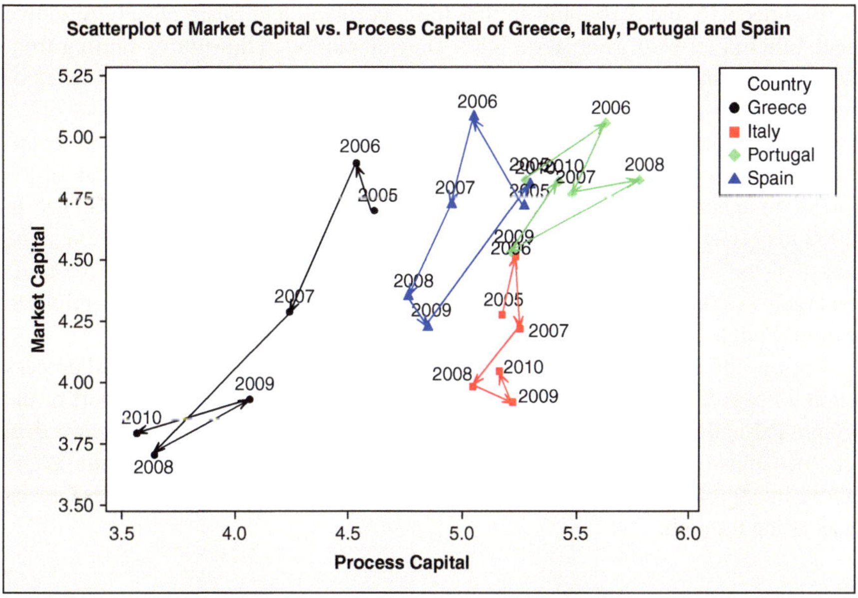

Fig. 3.15 Scatterplot of market capital versus process capital for Greece, Italy, Portugal, and Spain

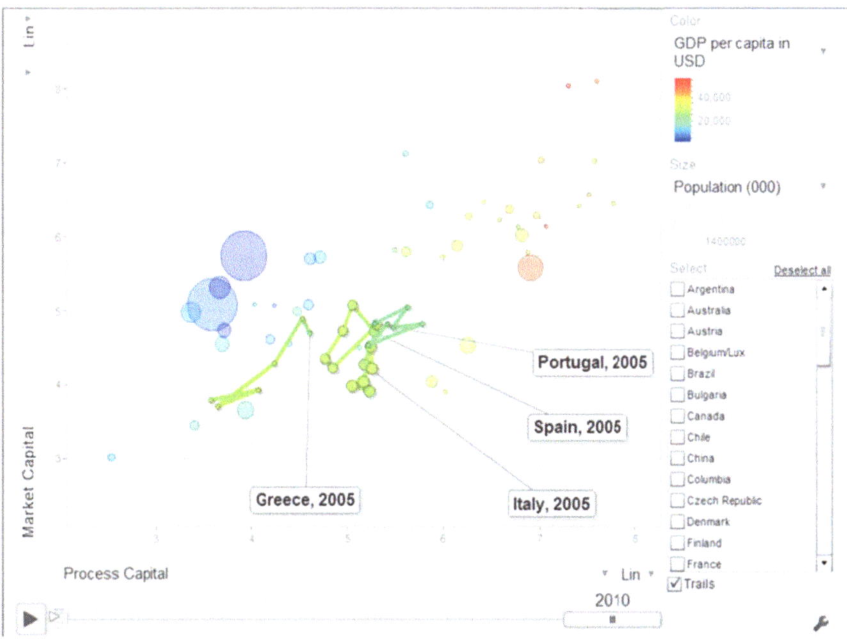

Fig. 3.16 Market capital versus process capital for the four Southern European countries

Figures 3.15 and 3.16 indicate that Italy, Portugal, and Spain closely resemble each other in terms of short-term NIC. They all exhibited downturns starting from 2006, yet bounced a little back in 2010. The short-term intangible assets development of Greece was a clear watershed, reflecting the more serious conditions of the Greek economic status even before the financial crisis. Actually, for all four countries, the signs of trouble surfaced in 2006—much earlier than the outbreak of the global financial crisis in September 2008. The temporary rebound of Greece in 2009 may reflect the effect of its stimulus package and less serious condition at the onset of the crisis reported in Chap. 2. However, after the sovereign debt crisis emerged in 2010, Greek short-term NIC again declined, as it was susceptible to external impact.

Figure 3.16 shows that, in a 48-country landscape, the market capital and process capital codevelopment of these four countries can be seen in the lower part of the graph. Particularly, Greek's short-term NIC had regressed to a level even lower than that of China and India's (the two large bubbles with higher market capital). The above intangible assets findings are in agreement with the economic troubles Greece was going through.

Dynamics of National Intellectual Capital in Three Time Periods

The previous illustrations and graphs describe more internal progression or regression of NIC in each country with individual capital score changes over the 6 years and path of development of various capitals. The ranking changes to be introduced in this section represent external competitiveness. Ranking, by and large, is a zero-sum game in which the advance of one country simultaneously marks the decline of another.

In order to probe the dynamics of NIC change before and after the financial crisis, we plotted Figs. 3.17, 3.18, 3.19, 3.20, 3.21, 3.22, 3.23, and 3.24 and Table 3.2 in three time periods: 2005–2006, 2007–2008, and 2009–2010. To have a clearer presentation, we separate the data in two sets. The first set contains capital scores and ranking changes of human capital, market capital, and process capital in Figs. 3.17, 3.19, 3.21, and 3.23. The second set exhibits capital scores and ranking changes of renewal capital, financial capital, and average NIC in Figs. 3.18, 3.20, 3.22, and 3.24.

The figures are mapped based on the mean capital scores and ranking changes of each period. Ranking change is next to each capital score and copied from the bold-faced ranking difference of column 2, 5, 7, and 9 of Table 3.2. Taking human capital ranking difference of Italy (−1, 3, −4, 0) as example, ranking change "−1" representing the difference between the 3rd (2009–2010) and 1st (2005–2006) period is plotted at year 2005; ranking change "3" representing the difference between the 2nd period (2007–2008) and 1st (2005–2006) is plotted at year 2007; ranking change "−4" representing the difference between the 3rd period (2009–

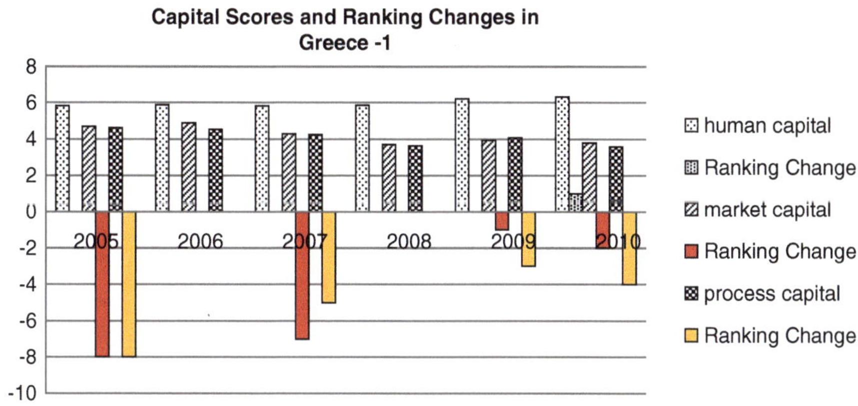

Fig. 3.17 Human capital, market capital, process capital, and ranking changes in Greece

2010) and 2nd (2007–2008) is plotted at year 2009; and ranking change "0" representing the difference between the 3rd period (2009–2010) and 6-year average ranking (exhibited in Table 3.1) is plotted at year 2010. Special attention can be paid to the negative numbers, which means that the level of a particular capital was higher before than after the financial crisis. Since it is easier to read figures than tables, we will describe the finding of each figure separately.

From Fig. 3.17 and Table 3.2, it is apparent that Greece's human capital does not change much over the three time periods, with one rank gain after the financial crisis, when comparing 2009–2010 with the 6-year average ranking. The negative numbers (−8, −7, -1, −2) for market capital indicate that Greece was losing international competitiveness in market capital after the financial crisis. The negative numbers (−8, −5, −3, −4) of process capital also show that Greece lost international competitiveness in process capital after the financial crisis. The relatively large–scale declines plotted at year 2005 and year 2007 show that Greece's market capital and process capital were comparatively strong in 2005–2006 before the financial crisis. Yet, they declined and lost competitiveness with the unfolding of the crisis.

Figure 3.18 and Table 3.2 show that the ranking changes of renewal capital, financial capital, and average NIC in Greece were relatively minor, ranging from −2 to +1. Yet, its decline of two rankings (−2) in renewal capital plotted at year 2005 and year 2009 deserves some attention, as renewal capability will be associated with Greece's future recovery.

Figure 3.19 and Table 3.2 show the ranking changes of human capital (−1, 3, −4, 0), market capital (−3, −1, −2, 0), and process capital (1, 4, −3, 0) for Italy. The negative numbers represent a loss of international competitiveness during different periods of financial crisis. Yet, the last ranking difference of "0" indicates in 2009–2010 Italy's human capital, market capital, and process capital regained its respective 6-year average ranking. In other words, after the financial crisis, these three capitals have regained its original international competitiveness.

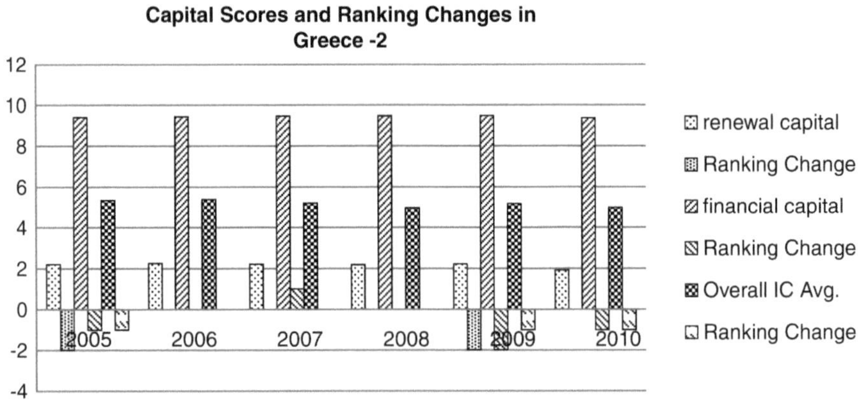

Fig. 3.18 Renewal capital, financial capital, average NIC, and ranking changes in Greece

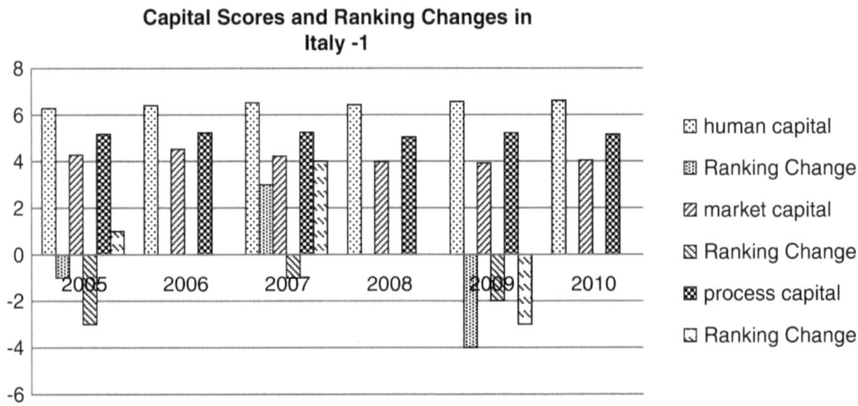

Fig. 3.19 Human capital, market capital, process capital, and ranking changes in Italy

Figure 3.20 and Table 3.2 show that Italy had minor ranking changes in terms of renewal capital, financial capital, and average NIC, ranging from −2 to +1. The ranking loss for renewal capital represents a loss of international competitiveness after the financial crisis.

Figure 3.21 and Table 3.2 show the ranking changes of human capital (−6, 2, −8, −4), market capital (−2, 2, −4, 1), and process capital (−1, 4, −5, −2) for Portugal. Portugal was losing human capital international competitiveness in a relatively large scale during and after the financial crisis, plotted in the figure at year 2005, 2009 and 2010, respectively. We particularly used dotted circles on Fig. 3.21 to highlight a matching phenomena found in Portugal during the financial crisis. The positive ranking gains highlighted with a blue dotted circle on the left side reflect the fact that Portugal was in relatively good shape at the onset of the financial crisis (2008). However, the debt problem which gradually emerged in 2010 (ranking loss highlighted with a red-dotted circle on the right side) resulted in Portugal's request for

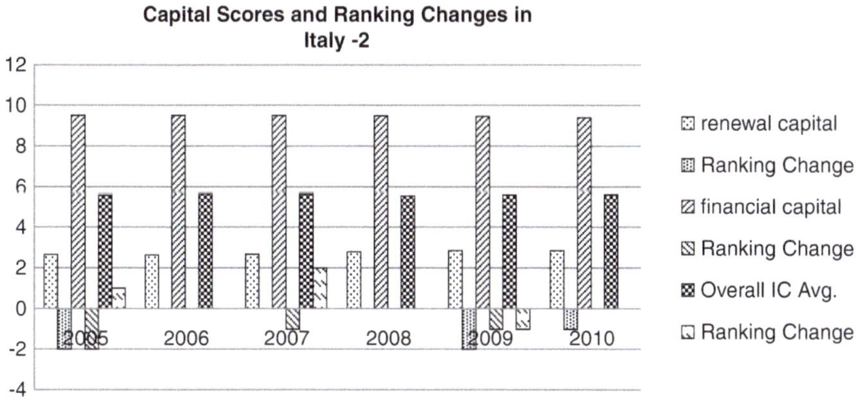

Fig. 3.20 Renewal capital, financial capital, average NIC, and ranking changes in Italy

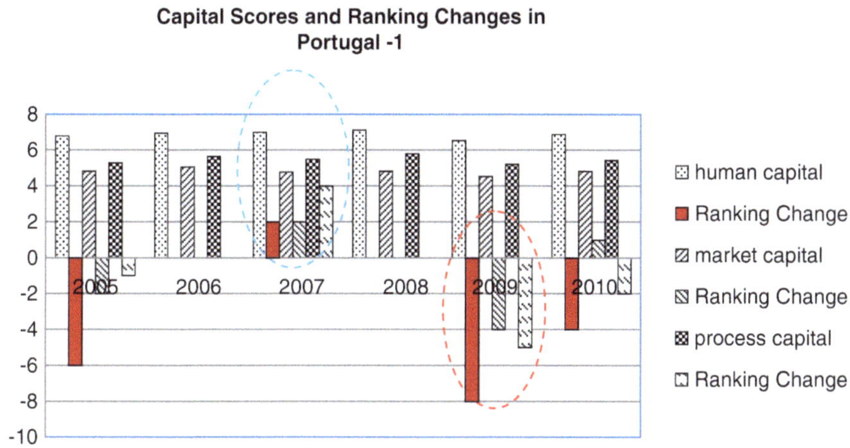

Fig. 3.21 Human capital, market capital, process capital, and ranking changes in Portugal

the April 2011 bailout. The concurrent ranking declines of human capital (−8), market capital (−4), and process capital (−5) plotted at year 2009 in the figure are the comparison of ranking changes between 2009–2010 and 2007–2008.

Figure 3.22 and Table 3.2 indicate that Portugal had impressive ranking gains (9, 7, 2, 2) in renewal capital, which were likely due to increasing investments in R&D. At the same time, Portugal's financial capital ranking changes were minor, ranging from −1 to +1. Since the last NIC comparison plotted at year 2010 in the figure shows no change, it means Portugal's most recent 2009–2010 NIC international competitiveness resumed to its original 6-year average.

Figure 3.23 and Table 3.2 show the ranking changes of human capital (2, 1, 1, 3), market capital (−3, −2, −1, 0), and process capital (0, 0, 0, 1) for Spain. All ranking gains in human capital indicate that Spain was gaining human capital international competitiveness during and after the financial crisis. Spain's market capital ranking

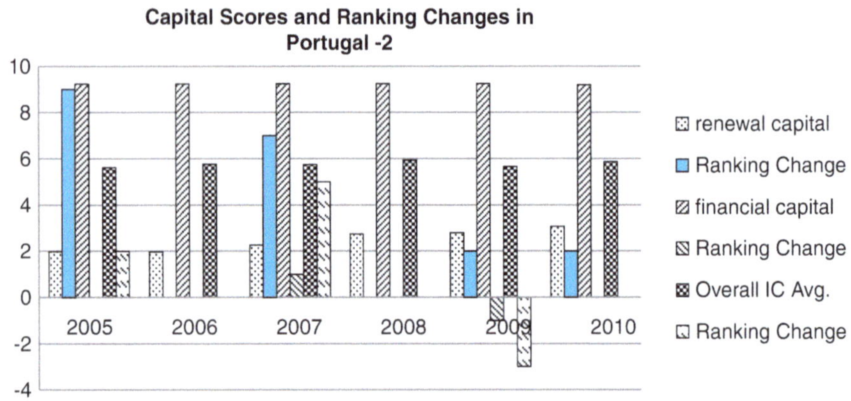

Fig. 3.22 Renewal capital, financial capital, average NIC, and ranking changes in Portugal

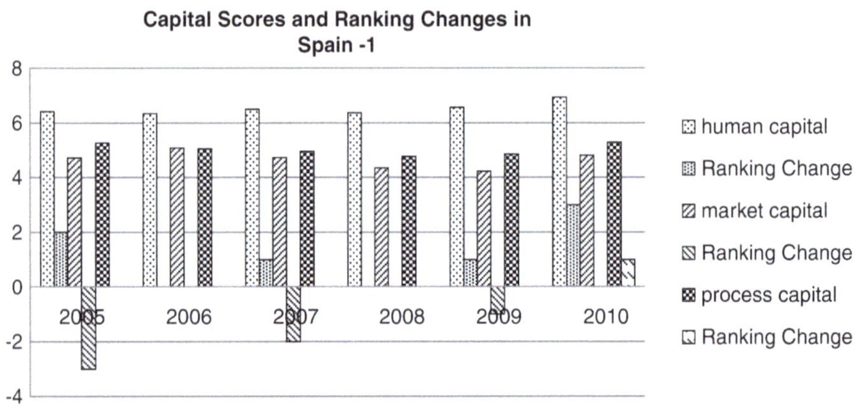

Fig. 3.23 Human capital, market capital, process capital, and ranking changes in Spain

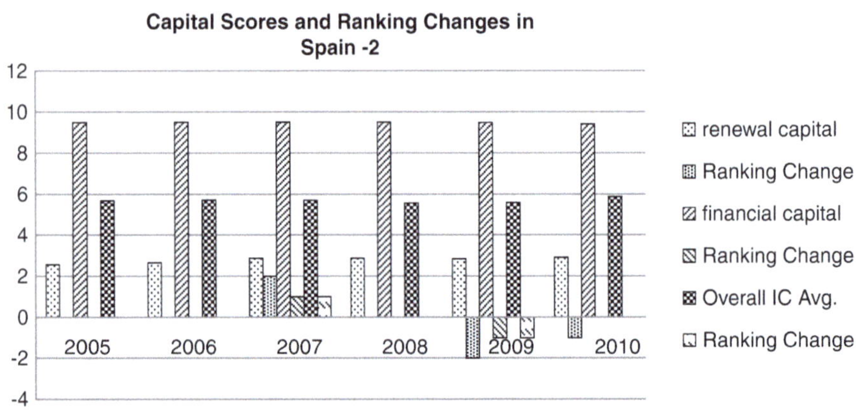

Fig. 3.24 Renewal capital, financial capital, average NIC, and ranking changes in Spain

Table 3.2 Ranking changes in three time periods for Greece, Italy, Portugal, and Spain

	(1st period vs. 3rd period)	1st period 2005–2006	2nd period 2007–2008		3rd period 2009–2010		2005–2010 average NIC48 ranking	
Human capital	Difference#1	Ranking	Ranking	Difference	Ranking	Difference	NIC 48 Ranking	Difference
Greece	**0**	30	30	**0**	30	**0**	31	**1**
Italy	**–1**	27	24	**3**	28	**–4**	28	**0**
Portugal	**–6**	20	18	**2**	26	**–8**	22	**–4**
Spain	**2**	26	25	**1**	24	**1**	27	**3**
Market Capital	Difference	Ranking	Ranking	Difference	Ranking	Difference	NIC 48 Ranking	Difference
Greece	**–8**	37	44	**–7**	45	**–1**	43	**–2**
Italy	**–3**	41	42	**–1**	44	**–2**	44	**0**
Portugal	**–2**	32	30	**2**	34	**–4**	35	**1**
Spain	**–3**	34	36	**–2**	37	**–1**	37	**0**
Process capital	Difference	Ranking	Ranking	Difference	Ranking	Difference	NIC 48 Ranking	Difference
Greece	**–8**	31	36	**–5**	39	**–3**	35	**–4**
Italy	**1**	28	24	**4**	27	**–3**	27	**0**
Portugal	**–1**	25	21	**4**	26	**–5**	24	**–2**
Spain	**0**	29	29	**0**	29	**0**	30	**1**
Renewal capital	Difference	Ranking	Ranking	Difference	Ranking	Difference	NIC 48 Ranking	Difference
Greece	**–2**	30	30	**0**	32	**–2**	32	**0**
Italy	**–2**	25	25	**0**	27	**–2**	26	**–1**
Portugal	**9**	34	27	**7**	25	**2**	27	**2**
Spain	**0**	26	24	**2**	26	**–2**	25	**–1**
Financial capital	Difference	Ranking	Ranking	Difference	Ranking	Difference	NIC 48 Ranking	Difference
Greece	**–1**	23	22	**1**	24	**–2**	23	**–1**
Italy	**–2**	20	21	**–1**	22	**–1**	22	**0**
Portugal	**0**	28	27	**1**	28	**–1**	28	**0**
Spain	**0**	21	20	**1**	21	**–1**	21	**0**

(continued)

Table 3.2 (continued)

	(1st period vs. 3rd period)	1st period 2005–2006	2nd period 2007–2008		3rd period 2009–2010		2005–2010 average NIC48 ranking	
NIC	Difference	Ranking	Ranking	Difference	Ranking	Difference	NIC 48 Ranking	Difference
Greece	**–1**	31	31	**0**	32	**–1**	31	**–1**
Italy	**1**	29	27	**2**	28	**–1**	28	**0**
Portugal	**2**	28	23	**5**	26	**–3**	26	**0**
Spain	**0**	27	26	**1**	27	**–1**	27	**0**

#Remarks:
1. The second column compares the 3rd (2009–2010) period ranking with the 1st (2005–2006)
2. The fifth column compares the 2nd (2007–2008) period ranking with the 1st (2005–2006)
3. The seventh column compares the 3rd (2009–2010) period ranking with the 2nd (2007–2008)
4. The last column compares the most current 3rd period ranking (2009–2010) with the original 6-year average NIC48 ranking summarized in Table 3.1

decline (−3, −2, −1) shows that it was losing market capital international competitiveness during the financial crisis. However, after the financial crisis (2009–2010), the ranking regained its 6-year average. Process capital for Spain did not experience changes in ranking during the financial crisis.

Figure 3.24 and Table 3.2 indicate that Spain had minor ranking changes in renewal capital, financial capital, and average NIC, ranging from −2 to +2. Yet, one rank loss of renewal capital, plotted at year 2010, deserves some attention as Spain was losing its renewal capital international competitiveness after the financial crisis when many countries were improving innovation and R&D.

In summary, within this country cluster, human capital for Italy and Portugal lost international competitiveness after the financial crisis, with Portugal experiencing a larger decline. Greece and Portugal had relatively large-scale market capital, and process capital declines. For renewal capital, 1–10 scaled financial capital and average NIC, most of the rankings after the financial crisis (2009–2010) regained their 6-year average ranking. By eyeballing Figs. 3.17– 3.24, Greece had the largest number of ranking declines and Portugal had the most impressive ranking gains in renewal capital.

An important finding of this section is that internal improvement does not guarantee external international competitiveness. An example is Portugal's process capital. Over the 6 years, its score advanced 0.15 point (5.43–5.28 in Fig. 3.3), yet it declined one rank for the same period. Therefore, a country's internal improvement and external competitiveness need to be examined together to get a fuller picture of its relative position in the international arena.

3-Dimensional National Intellectual Capital Trajectory

Up to this section, all the figures (Figs. 3.1–3.24) plotted different combinations of various capitals, capital-GDP codevelopment, and ranking changes on a flat plane. Even though Figs. 3.7, 3.8, 3.9, 3.10, 3.11, and 3.12 displayed four dimensions with color and bubble size as the 3rd and 4th dimension in addition to X-axis and Y-axis, graphically they are still on one plane. This section presents three-dimensional (3D) graphs showing each country's NIC development trajectory on a 48-country 3D landscape of various types. Since NIC evolution over time is always complex, combining different views from different angles may provide more comprehensive perspectives.

In what follows, we first explore the value-driving context by uncovering influencing factors of GDP growth for Greece, Italy, Portugal, and Spain from 2005 to 2010. Second, in explaining the GDP growth, we present the high, middle, and low capability which human capital, market capital, process capital, and renewal capital respectively have. Third, we examine the efficiency (GDP growth) enhancing factors and impeding factors of each country, taking each country's turning point into consideration. Last, we identify the efficiency drivers of each country in reaching a targeted GDP.

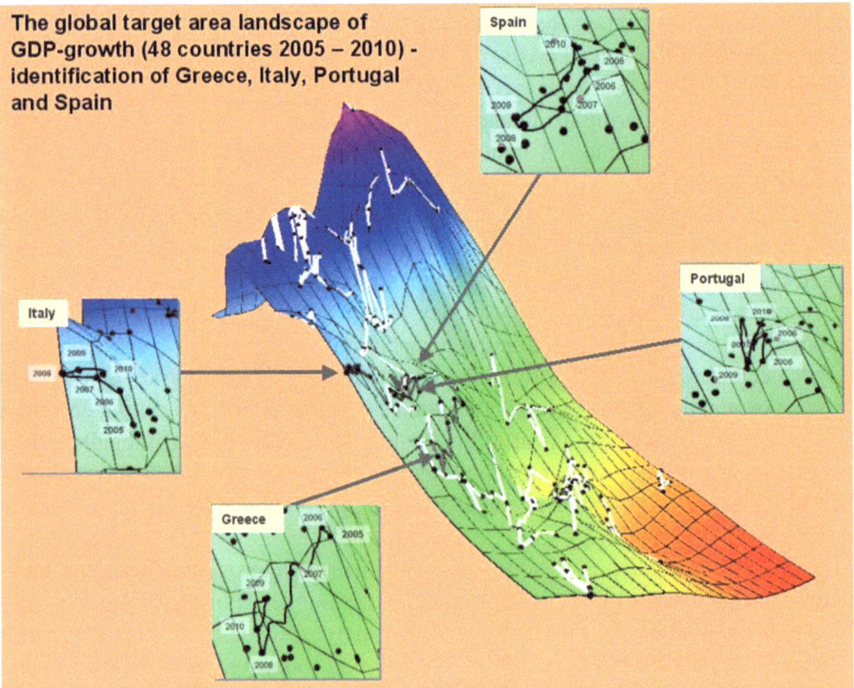

Fig. 3.25 The NIC trail of Greece, Italy, Portugal, and Spain on a 3D 48-country landscape

Figure 3.25 shows a rectangular 3D landscape, with a longer north–south side. The formation of the landscape is based on the data of 29 national intellectual capital indicators (presented in Appendix 3) for 48 countries covering 6 years (2005–2010), with a total of 8,352 (29x48x6) data points. In other words, Fig. 3.25 is the aggregated NIC landscape of the 48 countries. The relative NIC positions of Greece, Italy, Portugal, and Spain (pointed out with arrows) are in the middle left side of the landscape. The enlarged trails indicate the NIC progression (upward), regression (downward), or changing configuration (sideward) of each country over the 6 years. Each white trail on the landscape represents a country. The altitude coloring from purple, dark blue, blue, green, yellow, and red is computer generated, according to GDP values, with purple (north) representing high GDP descending to low GDP in red (south).

Figures 3.26, 3.27, and 3.28, are a set. Figure 3.26 presents four graphs indicating the high capability region of each human capital, market capital, process capital, and renewal capital in explaining the GDP growth. High capability means the data points are 20% above the mean value of each capital. This figure shows that none of the human capital, market capital, process capital, and renewal capital of these four Southern European countries are in the high capability region.

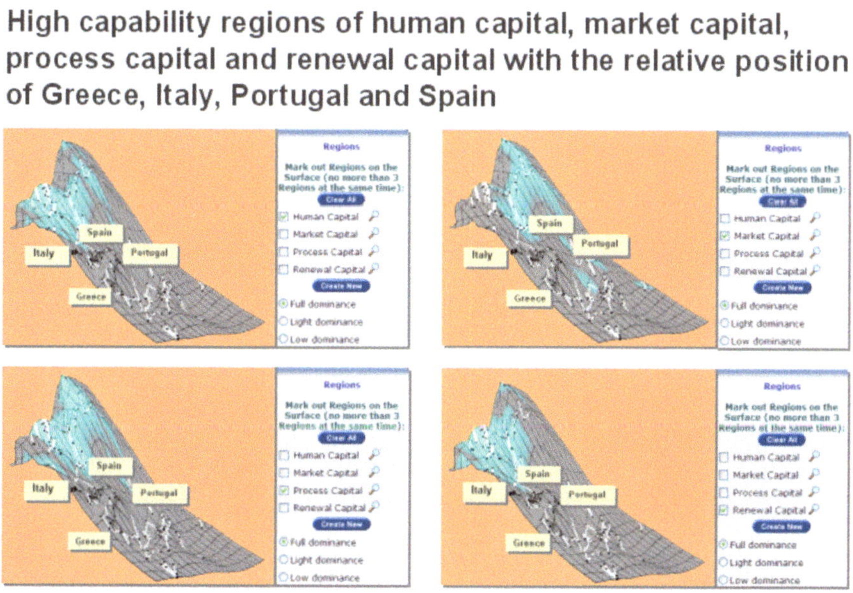

Fig. 3.26 The high capability region of human capital, market capital, process capital, and renewal capital

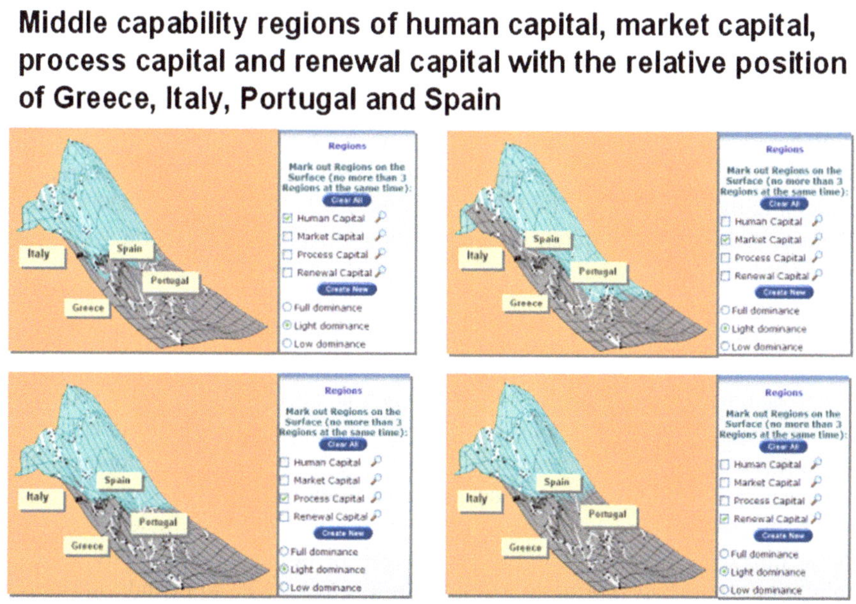

Fig. 3.27 The middle capability region of human capital, market capital, process capital, and renewal capital

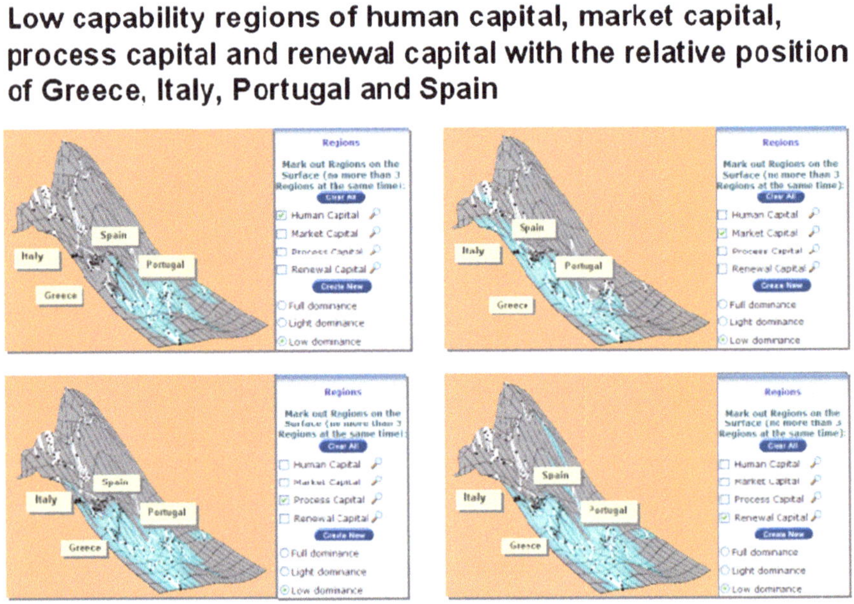

Fig. 3.28 The low capability region of human capital, market capital, process capital, and renewal capital

Figure 3.27 displays another four graphs indicating the middle capability region of each human capital, market capital, process capital, and renewal capital in explaining the GDP growth. Middle capability means the data points are within the top 50% value of each capital. This figure shows that human capital of Italy, Portugal, and Spain is positioned at the border area of the middle capability region.

Figure 3.28 shows the last four graphs in this set indicating the low capability region of each human capital, market capital, process capital, and renewal capital in explaining the GDP growth. Low capability means the data points are 20% below the mean value of each capital. This figure shows that most of the capitals of these four countries (except what described in Fig. 3.27) are in the low capability region.

In general, Figs. 3.26, 3.27, and 3.28 as a set explain that NIC of these four countries are generally 20% below average in explaining GDP growth. All these four Southern European countries are classified by the World Economic Forum as Stage 3 innovation-driven developed countries. Yet, finding that most of their NIC are 20% below average is a warning sign for relevant policy makers. In an era when intangible assets play an increasingly important role in national development, the messages revealed in these graphs deserve special attention.

Another set of figure, Figs. 3.29, 3.30, 3.31, and 3.32, display efficiency enhancing or impeding factors that affect GDP growth before and after the NIC turning point for Greece, Italy, Portugal, and Spain, respectively. The white indicator boxes at the upper right display the results of the enhancing or impeding factors. We employed three or four rounds of data analysis for each country, depending on

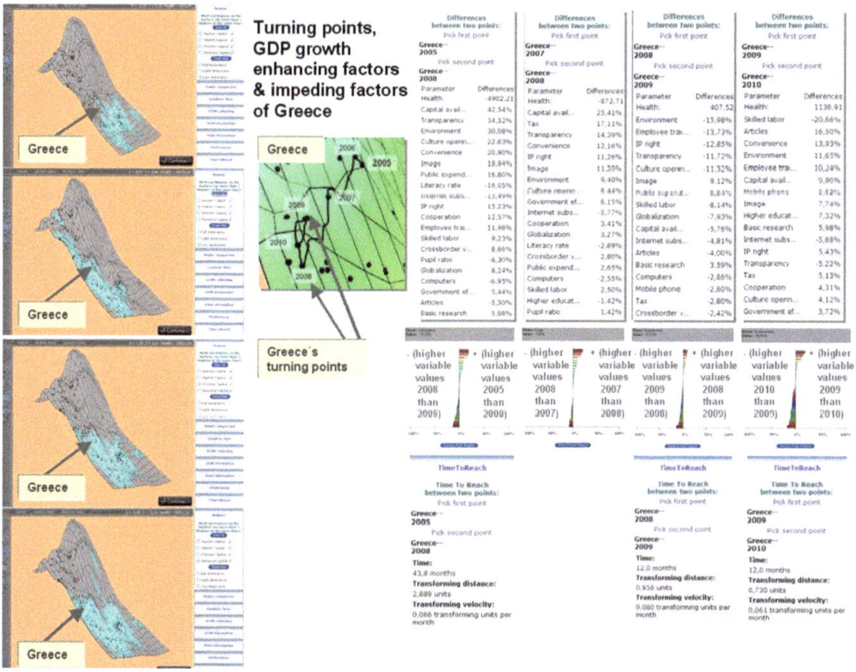

Fig. 3.29 Turning point and GDP growth enhancing and impeding factors of Greece

whether the country has one or two NIC turning points. For instance, Greece's turning points are 2008 and 2009; therefore, we traced the explaining factors of 2005 against 2008 (starting point → the first turning point), 2007 against 2008 (one year before the turning point → the first turning point), 2008 against 2009 (one year before the turning point → the second turning point), and 2009 against 2010 (the second turning point → the ending point).

As an easy reference, we summarize in Table 3.3 two sets of results, each with the first five key factors. If there is a clear percentage difference cutting point (exhibited in the white indicator boxes), the reported factors may be less than five. The first set displays the efficiency enhancing and impeding factors before the turning point, and the second set exhibits the efficiency enhancing and impeding indicators after the (2nd) turning point in italic letters. Since the factor strength (percentage) varies with different explaining power, we have calculated the average weight of the five enhancing and impeding factors and display each respective percentage in the rows for "Factor #5". We also count the frequency of four more frequently displayed categories, namely, government-related issues (G), human capital-related issues (H), market capital-related issues (M), and research and development issues (R&D).

Figure 3.29 shows the turning point (2008 and 2009), the enhancing and impeding factors of Greece. The four rectangular 3D graphs on the left-hand side of the figure indicate that all human capital, market capital, process capital, and renewal capital for Greece fall into the low capability region (20% below average).

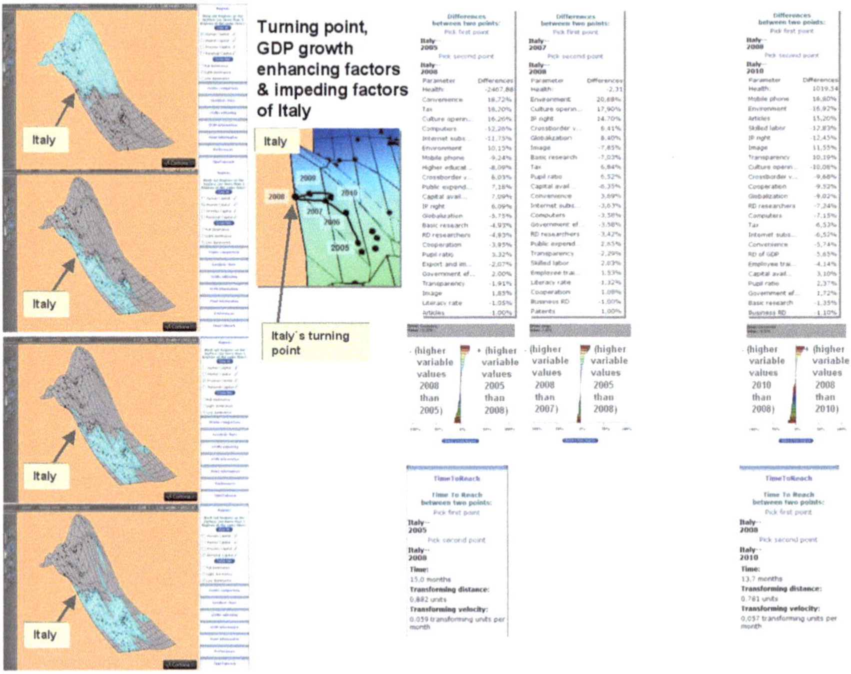

Fig. 3.30 Turning point and GDP growth enhancing and impeding factors of Italy

Table 3.3 shows that "capital availability" impeded Greece GDP growth both before and after the turning points, which reflects its serious financial troubles after the financial crisis. There are two more important warnings for reducing the hindrance to Greece GDP growth. First, "convenience of establishing new firms" is in both the impeding list of before and after the turning point; second, "scientific articles" changed from an enhancing factor before the turning point to an impeding factor after the turning point. Yet, "transparency of government policy" improved from an impeding factor before the turning point to an enhancing factor after the turning point. The average weights of impeding factors for both before and after the turning point are relatively large (30.13% and 12.44%). Therefore, other impeding factors, such as "fair business competition environment," "openness of culture," "mobile phone subscribers," and "image of country," also deserve some attention for future improvement.

In Fig. 3.30, Italy's turning point is 2008. The four rectangular 3D graphs on the left-hand side indicate that human capital of Italy is positioned at the border of the middle capability region (top 50%) and market capital, process capital, and renewal capital all fall into the low capability region (20% below average). Table 3.3 shows that "capital availability" impedes Italy GDP growth after the turning point, which also reflects its financial problem after the financial crisis, although less pressing than that of Greece or Portugal. In addition, "mobile phone subscribers" degraded from an enhancing factor before the turning point to becoming an impeding factor after the turning point. Yet, three improvements can be found in Italy: "fair business

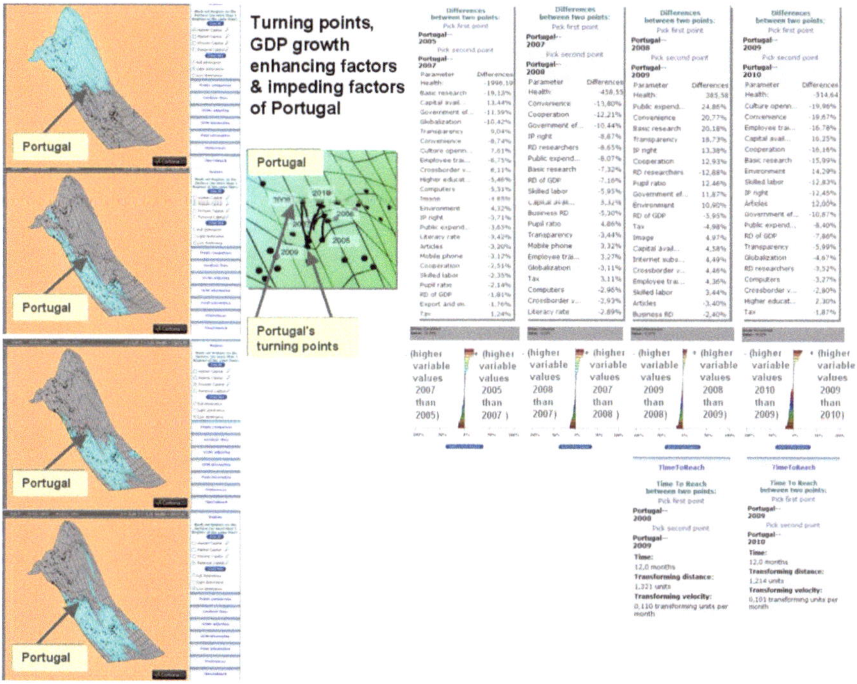

Fig. 3.31 Turning point and GDP growth enhancing and impeding factors of Portugal

competition environment," "openness of culture," and "cross-border ventures" all changed from impeding factors before the turning point to enhancing factors after the turning point. Other impeding factors, such as "convenience of establishing new firms," "corporate tax encouragement," "scientific articles," "image of country," and "transparency of government policy," indicate areas for attention to achieve better GDP growth.

Figure 3.31 shows two turning points for Portugal, 2008 and 2009. The four rectangular 3D graphs on the left-hand side of Fig. 3.31 indicate that human capital of Portugal is positioned at the border of the middle capability region (top 50%) and that of market capital, process capital, and renewal capital all fall into the low capability region (20% below average). Table 3.3 shows that "capital availability" impedes Portugal GDP growth both before and after the turning points and both sequenced as number one, which truly reflects its serious financial troubles during and after the financial crisis. Yet, there are four good signs for Portugal. The first three are "basic research," "convenience of establishing new firms," and "university-enterprise cooperation." These were enhancing factors before and after the turning points. The fourth is "openness of culture," which changed from an impeding factor before the turning point to an enhancing factor after the turning point. Other impeding factors, such as "fair business competition environment," "transparency of government policy," "scientific articles," "corporate tax encouragement," and "higher education enrollment," indicate areas for improvement to achieve better GDP growth.

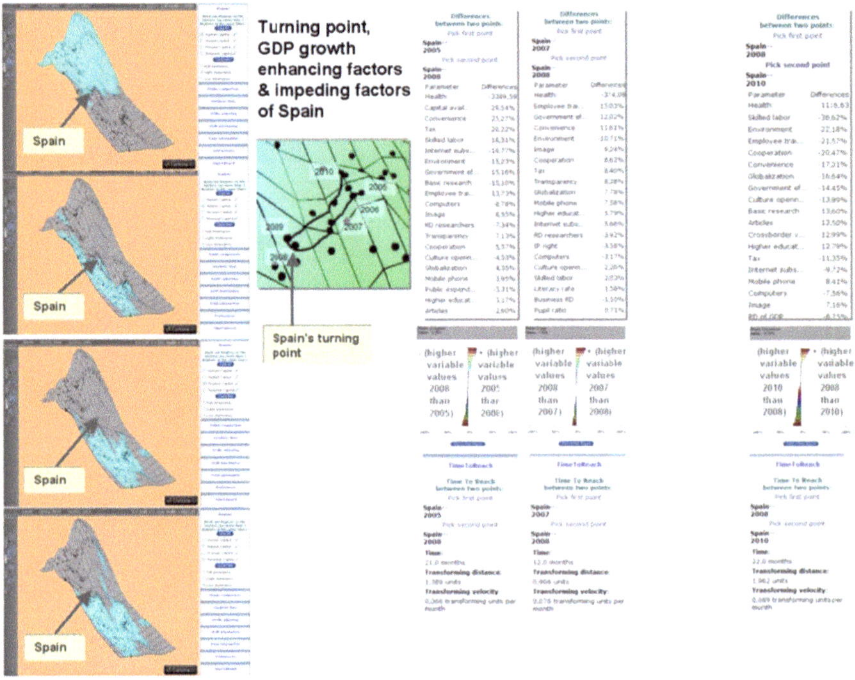

Fig. 3.32 Turning point and GDP growth enhancing and impeding factors of Spain

Figure 3.32 shows that 2008 was Spain's turning point. The four rectangular 3D graphs on the left-hand side of Fig. 3.32 indicate that the human capital of Spain is positioned at the border of the middle capability region (top 50%) and market capital, process capital, and renewal capital all fall into the low capability region (20% below average). Table 3.3 shows that "capital availability" impedes Spain GDP growth before the turning points, which reflects its financial problem during the financial crisis, albeit less pressing than that of Greece or Portugal. In addition, its "basic research" degraded from an enhancing factor before the turning point to an impeding factor after the turning point. Yet, there are three good signs. The first one is "fair business competition environment" remained an enhancing factor both before and after the turning point. The second and third ones "skilled labor" and "convenience of establishing new firms" changed from an impeding factor before the turning point to an enhancing factor after the turning point. Since the average weight of impeding factors before the turning point is relatively large (21.70%), other impeding factors, such as "scientific articles," "corporate tax encouragement," "higher education enrollment," "mobile phone subscribers," "government efficiency," and "image of country," also deserve some attention for future improvement.

To summarize Table 3.3 into an overall suggestion, we count the frequency of both enhancing and impeding factors by categories, shown in the last row of the table. Government-related issues (G & G-C) have a total of 29 counts. Since capital availability is a concern for these four countries, we particularly highlighted it by

Table 3.3 Enhancing factors and impeding factors of GDP growth for Greece, Italy, Portugal, and Spain

Influencing factor	Greece Enhancing factors	Greece Impeding factors	Italy Enhancing factors	Italy Impeding factors	Portugal Enhancing factors	Portugal Impeding factors	Spain Enhancing factors	Spain Impeding factors
Factor #1 before (1st) turning point	Literacy rate (H)	Capital availability (G-C)	Computers in use per capita	Convenience of establishing new firms (G)	Basic research (R&D)	Capital availability (G-C)	Internet subscribers	Capital availability (G-C)
Factor#1 after (2nd) turning point	*Skilled labor* (H)	*Scientific articles* (R&D)	*Fair business competition environment* (G)	*Mobile phone subscribers*	*Openness of culture* (M)	*Capital availability* (G-C)	*Skilled labor* (H)	*Basic research* (R&D)
Factor #2 before (1st) turning point	Internet subscribers	Transparency of government policy (G)	Internet subscribers	Corporate tax encouragement (G)	Government efficiency (G)	Transparency of government policy (G)	Fair business competition environment (G)	Convenience of establishing new firms (G)
Factor #2 after (2nd) turning point	*Internet subscribers*	*Convenience of establishing new firms* (G)	*Skilled labor* (H)	*Scientific articles*(R&D)	*Convenience of establishing new firms* (G)	*Fair business competition environment* (G)	*Fair business competition environment* (G)	*Scientific articles*(R&D)
Factor #3 before (1st) turning point	Pupil-teacher ratio (H)	Fair business competition environment (G)	Mobile phone subscribers	Openness of culture (M)	Globalization (M)	Openness of culture (M)	Basic research (R&D)	Corporate tax encouragement (G)
Factor #3 after (2nd) turning point	*Transparency of government policy* (G)	*Fair business competition environment* (G)	*IP rights protection* (R&D)	*Image of country* (M)	*Employee training* (H)	*Scientific articles*(R&D)	*Employee training* (H)	*Higher education enrollment*(H)
Factor #4 before (1st) turning point	Computers in use per capita	Openness of culture (M)	Higher education enrollment (H)	Fair business competition environment (G)	Convenience of establishing new firms (G)	Cross-border ventures (M)	Computers in use per capita	Skilled labor(H)

(continued)

Table 3.3 (continued)

Influencing factor	Greece Enhancing factors	Greece Impeding factors	Italy Enhancing factors	Italy Impeding factors	Portugal Enhancing factors	Portugal Impeding factors	Spain Enhancing factors	Spain Impeding factors
Factor #4 after (2nd) turning point	–	*Employee training* (H)	*Openness of culture* (M)	*Transparency of government policy* (G)	*University-ent. cooperation* (R&D)	*Higher education enrollment* (H)	*University-enterprise cooperation*	*Mobile subscribers*
Factor #5 before (1st) turning point	Scientific articles (R&D)	Convenience of establishing new firms (G)	Globalization (M)	Cross border ventures (M)	Employee training (H)	Fair business competition environment (G)	R&D researchers (R&D)	Government efficiency (G)
	*(Average weight – 10.42%)*a	*(Average weight – 30.13%)*a	*(Average weight – 9.42%)*	*(Average weight – 14.27%)*	*(Average weight – 11.33%)*	*(Average weight – 8.10%)*	*(Average weight – 12.64%)*	*(Average weight – 21.70%)*
Factor #5 after (2nd) turning point	–	*Capital availability* (G-C)	*Cross border ventures* (M)	*Capital availability* (G-C)	*Basic research* (R&D)	–	*Convenience of establishing new firms* (G)	*Image of country* (M)
	(Average weight – 10.59%)	*(Average weight – 12.44%)*	*(Average weight – 12.39%)*	*(Average weight – 11.77%)*	*(Average weight – 17.71%)*	*(Average weight – 11.21%)*	*(Average weight – 23.61%)*	*(Average weight – 11.09%)*
Category frequency count	G – 1 H – 3 M – 0 R&D – 1	G – 7 H – 1 M –1 R&D – 1	G – 1 H – 2 M – 3 R&D – 1	G –5 H – 0 M – 3 R&D – 1	G – 3 H – 2 M – 2 R&D – 3	G – 5 H – 1 M – 2 R&D – 1	G – 3 H – 2 M –0 R&D – 2	G – 4 H – 2 M – 1 R&D – 2

Remarks: (G & G-C) – government related issues including capital availability represented by G-C (total 29 counts including 6 counts of capital availability), (H) – human capital issues (13 counts), (M) – market capital issues (12 counts), (R&D) – research and development (12 counts)
#: average weight of each enhancing/impeding factor before and after the turning point

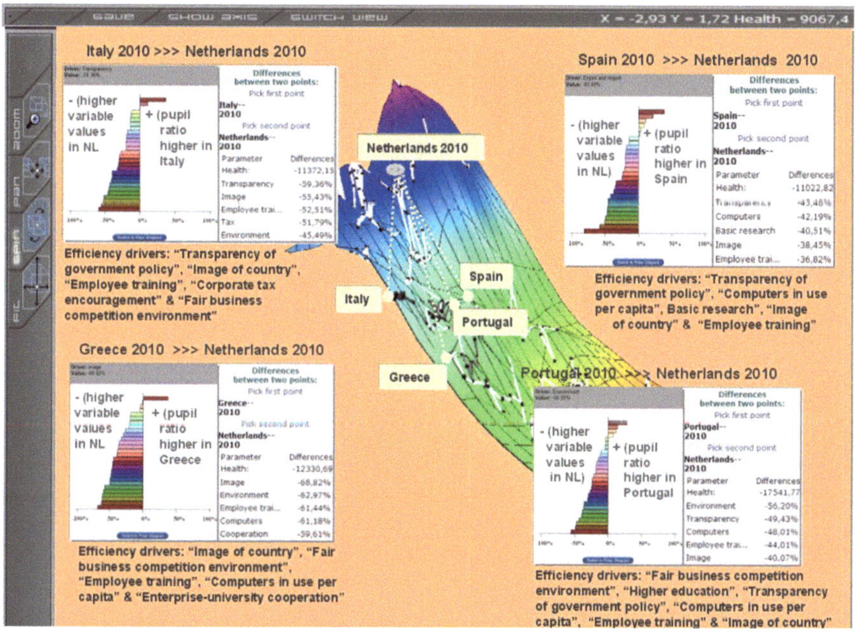

Fig. 3.33 Efficiency drivers and distance to targeted GDP of the Netherlands

G-C. Out of the 29 counts, 6 counts are capital availability with two each for Greece and Portugal and one each for Italy and Spain. This result truly reflects what had happened during this financial turmoil. Among the 29 counts, 21 of them are impeding factors including convenience of establishing new firms, corporate tax encouragement, fair business competition environment, government efficiency, and transparency of government policy. Human capital (H)-related issues have 13 counts, market capital (M)-related issues have 12 counts, and research and development (R&D)-related issues have 12 counts. From Table 3.3, government-related issues turn out to be the greatest barricade to achieve GDP growth. Therefore, governments of these four countries need to launch aggressive structural reforms to alleviate relevant impediments for facilitating future economic growth.

Figure 3.33 identifies the efficiency drivers (positive term) for each individual country. Efficiency targeting is akin to benchmarking. Efficiency represents GDP in this 3D software program. Since the Netherlands is an old time ocean superpower in Europe similar to Portugal and Spain and is still very competitive in the contemporary world, it serves as a good benchmarking target for these four Southern European countries. In Fig. 3.33, the stacked small horizontal stripes are the deficiency factors of each country benchmarked against the GDP of the Netherlands in 2010. The length of the stripes represents the degree of deficiency. For easier comparison, we list the first five deficiency factors of each country in Table 3.4.

Table 3.4 The first five efficiency drivers targeting GDP of the Netherlands

Efficiency driver	Greece	Italy	Portugal	Spain
Factor #1	Image of country	Transparency of government policy	Fair business competition environment	Transparency of government policy
Factor #2	Fair business competition environment	Image of country	Transparency of government policy	Computers in use per capita
Factor #3	Employee training	Employee training	Computers in use per capita	Basic research
Factor #4	Computers in use per capita	Corporate tax encouragement	Employee training	Image of country
Factor #5	University-enterprise cooperation	Fair business competition environment	Image of country	Employee training
Average deficiency percentage of the above 5 drivers	–62.80%	–52.92%	–47.54%	–40.69%

The results indicate that Greece needs to strengthen, in sequence, its "image of country," "fair business competition environment," "employee training," "computers in use per capita," and "university-enterprise cooperation." Italy needs to strengthen its "transparency of government policy," "image of country," "employee training," "corporate tax encouragement," and "fair business competition environment." Portugal needs to strengthen its "fair business competition environment," "transparency of government policy," "computers in use per capita," "employee training," and "image of country." Spain needs to strengthen its "transparency of government policy," "computers in use per capita," "basic research," "image of country," and "employee training."

With the Netherlands being used as a benchmarking country, irrespective to the sequence of deficiency, the common areas (in frequency count) that these four countries need to improve include image of country (4 counts), employee training (4 counts), transparency of government policy (3 counts), fair business competition environment (3 counts), and computers in use per capita (3 counts). That is, the areas that need improvement are quite similar for these four countries. Among the above five common indicators, three (image of country, transparency of government policy, and fair business competition environment) require government's reforms and two (employee training and computers per capita) need resource input. Based on the average deficiency (negative term) percentage reported in Table 3.4 and Fig. 3.33, Greece (−62.80%) has the most amount of work to do in order to achieve the target GDP (Netherlands'), followed by Italy (−52.92%), Portugal (−47.54%), and Spain (−40.69%). In other words, Spain will have the shortest route to reach the preset GDP goal through enhancing transparency of government policy, investing more resources for people to utilize information technology, and upgrading basic research and so on.

The added value of the above 3D trajectory analysis is to (1) display the NIC trails of the discussed countries on a 48-country landscape; (2) locate the capability region of the four individual capitals for each country; (3) highlight NIC turning points, details of the enhancing or impeding factors, and the dynamics of factor changes for each country; and (4) calculate the distance that each country has to progress in order to achieve the benchmark GDP and identify the first five key factors to reach this goal. This section supplements the previous sections by looking into the contribution of each individual indicator in each year, rather than average scores and average rankings.

In this chapter, we reported our research findings in layman terms with graphs and tables of various types for easier reading. This study on national intangible assets has uncovered some important findings that are not apparent in other GDP-related studies. For instance, the contribution of renewal capital to GDP growth is more effective in less-developed countries than in advanced countries; the distance to the target GDP for Italy (52.92%) is farther than that of Portugal (47.54%) even though Italy did not request for a bailout after the financial crisis.

Chapter 4
Beyond the 2008 Global Financial Crisis

Since early 2010, the Euro zone has been facing a major debt crisis. The governments of several countries in the zone, including Greece and Portugal, accumulated what many consider to be unsustainable levels of government debt. Gradually, the crisis threatened to spread to Italy and Spain, the third and fourth largest economies in the Euro zone, respectively. The sovereign debt crisis developed as countries experienced higher deficits and growing debt, making governments increasingly unable to pay back the money they owed. Eventually, credit-rating agencies downgraded the sovereign bonds of several European countries, which made issuing new bonds extremely expensive. Consequently, Greece and Portugal had to turn to other European countries, European Central Bank, and the International Monetary Fund (IMF) for loans in order to avoid defaulting on their debt.

With the unfolding of the debt crisis, the entire EU was taking steps to solve the problem, while simultaneously preventing another crisis from occurring. In May 2010, EU member states agreed upon the establishment of a US$633.6 billion (€440 billion) Euro zone Member State-guaranteed European Financial Stability Facility (EFSF), after bailing out Greece (Moloney 2010). The EFSF will provide loans to EU member states in financial trouble. If a country receives a loan from the EFSF, it will be subject to strict austerity measures such as tax and pension reform, decreases in public wages, and privatization of some industries. EU member states will guaranty the EFSF's bonds. Yet, the EFSF is scheduled to expire in 2013, at which time it will be replaced with a permanent crisis mechanism called the European Stabilization Mechanism ("ESM"). To avoid any future crises in the Euro zone, the ESM will provide loans to a member state that is threatened with severe economic difficulties and will require borrowers to follow austerity measures similar to those of the EFSF (Harrington 2011).

With years of recession, it is clear that there were many contributing factors to the Euro zone debt crisis. Among them, the lack of supervision and oversight is widely criticized. To ensure fiscal discipline after the introduction of the Euro, the Stability and Growth Pact ("the Pact") was already stipulated in 1997. The Pact required that countries maintain a government budget deficit of no more than 3% of the GDP and a country's public debt cannot exceed 60% of the GDP. The European

C.Y.-Y. Lin et al., *National Intellectual Capital and the Financial Crisis in Greece, Italy, Portugal, and Spain*, SpringerBriefs in Economics 7,
DOI 10.1007/978-1-4614-5990-3_4, © The Author(s) 2013

Council (one of the EU's regulatory bodies) has the authority to penalize any member that does not abide by these restrictions. Nevertheless, there are no fixed rules governing penalties. Although the Pact sets guidelines, there is no oversight of fiscal policies and nations are left to self-regulate. Analysts argue that the lack of enforcement of the Pact was partially responsible for the Euro sovereign debt crisis (Moloney 2010). For example, Fig. 2.2 indicates that government debt in Greece, Italy and Portugal already exceeded 60% of its GDP as early as 2005. It seems that no penalty was enforced to avoid this wide spreading debt crisis from happening.

As a result, the EU had to bailout Greece and Portugal from damaging the economic stability of other Euro zone countries. The bailouts were designed to tide both countries over until they could afford to borrow commercially again. Spain's and Italy's economies are far bigger than those of Greece and Portugal and the European Union would try everything possible to prevent bailout requests from them (BBC 2012b). The fiscal problems of these four Southern European countries had already slowed down Euro zone's recovery from the 2008 global financial crisis, while other parts of the world were making good progress in that regard. In retrospect, Moloney (2010) commented that:

> At the core of the EU crisis was a destructive imbalance in the regulatory and supervisory architecture. The regulatory structure facilitated the cross-border activities…, but it did not adequately address cross-border supervision, co-ordination, crisis resolution, and deposit protection. There were no mechanisms for common decision-making in crisis. There was no system for macro-prudential supervision or early warnings.

In what follows, in the sequence of Greece, Italy, Portugal and Spain, we briefly describe what had happened in each individual country after 2010 with some prior reasons.

Greece

After the global financial crisis, the Greek economy contracted by 2% in 2009, and 4.8% in 2010; the budget deficit was cut by about 5% of the GDP in 2010 (CIA 2011). The strong fiscal contraction undercut domestic demand. As a result, output declined by over 4% in 2010 and the unemployment rate rose to almost 16% in early 2011. In January 2010, Greece's budget deficit in 2009 was revised upwards to 12.7% from 3.7%; that is more than four times the maximum allowed by EU rules. After rechecking Greece's accounts, the EU announced in April 2010 that the Greek deficit was even worse than thought – 13.6% of the GDP, not 12.7% (BBC 2012a).

The chain effect of the Greek financial crisis stems from a credibility gap caused by inaccurate and misreported statistics, as well as consistent underperformance on following through with reforms. This prompted major credit rating agencies in late 2009 to downgrade Greece's international debt rating. As a result, Greece was not able to repay debts with hiking interest rate in the commercial market.

After the debt crisis deteriorated, Greece took strong actions to deal with its most serious post-war economic crisis and embarked on an ambitious adjustment program to wade through the deep economic troubles by restoring sustainable public finances, competitiveness and the foundations for healthy and solid long-term growth. In early March 2010, the program contains the most severe austerity measures ever taken by the Greek government. It included an overall average 7–10% reduction of nominal public sector wages, cutting government spending (including pension and social service payments), raising the retirement age, reducing the size of the public sector, implementing new business taxes and higher VAT taxes, decreasing tax evasion, reforming the health care and pension systems, and improving competitiveness through structural reforms to the labor and product markets (Pagoulatos 2010; OECD 2011a). The economic adjustment plan was implemented with the technical and financial support of the IMF, the European Union, the European Central Bank, and the participation of Greece's private sector.

To prevent Greece from defaulting on its debts, in May 2010 Euro zone leaders and the IMF announced a 3-year package of US$158 billion (€110 billion) in loans to Greece at market-based interest rates for Greece to make debt repayments to creditors. Of the US$158 billion, Euro zone countries pledged to contribute US$115 billion (€80 billion) and the IMF pledged to contribute US$43 billion (€30 billion) (Nelson et al. 2011). An austerity program outlined in May 2010 aimed to reduce the government's budget deficit by 11 percentage points through 2013, bringing it below 3% of the GDP by 2014. In exchange for the largest bailout ever assembled, the Greek government announced combined spending cuts and tax increases totaling US$40 billion (€28 billion) over 3 years, on top of the tough austerity measures already taken. Greece struggled to boost revenues and cut spending to meet 2010 targets set by the EU and the IMF, especially after Eurostat – the EU's statistical office – revised upward Greece's deficit and debt numbers for 2009 and 2010 (CIA 2011).

With these measures in place, some promising signs have begun to show. Competitiveness gains and a rise in exports in late 2010 and early 2011 were the first signs of adjustment and reallocation. Excluding the effects of tax increases, inflation was close to zero in early 2011, 0.62% in April and 0.18% in May (Inflation.eu 2011). Despite moderately positive quarter-on-quarter GDP growth in the first quarter of 2011, the recession was still set to continue.

Unfortunately, market confidence kept on falling and Greece could not meet its ends. In July 2011, the IMF estimated that Greece's public debt increased substantially between 2010 and 2011, from 143% to 166% of the GDP. In the same month, European leaders announced a second likely bailout package for Greece, designed to resolve the Greek crisis and prevent contagion among other European economies (BBC 2012a; OECD 2011a). In return, a team of monitors would be based in Athens to ensure agreed reforms were put into place (BBC 2012b). This second financial assistance program provided loans to Greece on more favorable terms than the first package. The key provisions include reduced interest rates, longer maturities, a meaningful extension of maturities on existing Euro zone loans, some private-sector involvement, protection of Greek banks and measures to foster economic growth in

Greece (OECD 2011a). In addition, Greece had won a 53.5% reduction in its debt burden to private creditors, while any profits made by Euro zone central banks on their holdings of Greek debt will be channeled back to Greece (BBC 2012b).

In February 2012, the Euro zone cast doubt on the Greece's figures, requesting Athens to find a further US$427 million (€325 million) in budget cuts to get the second bailout (BBC 2012a). In early March 2012, Euro zone finance ministers approved US$172 billion (€130 billion) second bailout package for Greece to resolve the debt-ridden nation's immediate repayment needs (Al Jazeera and Agencies 2012).

Italy

In the initial stage of the global financial crisis, Italy was not seriously impacted. The problems of the American housing market had not affected Italy because its financial institutions did not own a large quantity of sub-prime bonds (Quirico 2010). Unfortunately, even though Italian banks did not participate in such types of lending that precipitated the crisis, they were caught up in the aftermath (Gurria 2011b). The real economic crisis confronting Italy was the decline in productivity over the last decade, which had spawned stagnating incomes, rising unit labor costs, a significant competitiveness gap, and tepid growth. Italy's huge public debt and corruption in government also contributed to the decline of its economy (Gurria 2011b).

Unlike most countries, Italy's response to the financial crisis and recession consisted of shifting expenditure towards social and some industrial support, rather than a deficit-increasing fiscal stimulus. The measures were designed to be budget-neutral, with some small stimulus from the redeployment of public spending towards categories considered to have higher multiplier effects. Also different from other countries, private debt in Italy was very low (Quirico 2010). In addition, Italian banks refused to lend money to each other, citing a lack of liquidity and an uncertainty about the financial soundness of borrowers.

Other than supporting banks and cutting public expenditures, universities and university reform are another chronic problem that Italian government has tried to solve since the start of the millennium. University policies, until the end of the 1990s, dramatically increased the number of professors and lecturers and, as a consequence, universities' costs (Quirico 2010). Over the years, universities' budgets worsened and risks of bankruptcy became real. Moreover, automatic wage increases every 2 years and the low thresholds for career advancement of many lecturers and professors made the budget more rigid and difficult to cut (Quirico 2010). Yet, reform of universities and the public administration seems mainly to have been oriented towards cutting costs rather than supporting efficiency, innovation and R&D.

In Italy, few industries fund university research and few industries or institutions have their own research sections. Consequently, companies without high-tech competitiveness either moved out of the country for cost-reduction, thus reducing job opportunities, or worked with foreign patents and relied upon foreign research activities. Such development has resulted in Italy's losing global competitiveness at a rather serious level (Quirico 2010).

With the continued financial problems and the result of the global economic recession, Italian unemployment rose from 6.2% in 2007 to 8.4% in 2011 (CIA 2012). Particularly, in the first two months of 2010, the youth unemployment rate in Italy rose to 28.2%, a growth of four percentage points compared to that of February 2009 (Eurofound 2010). The government also faces pressure from investors and European partners to address Italy's long-standing structural impediments to growth, such as an inflexible labor market and widespread tax evasion. In 2010, Italy raised the effective retirement age (linked to life expectancy) and has reduced the projected ratio of pension expenditure to GDP over the next 35 years; at the same time, it requires higher activity rates among older age groups (OECD 2011b). Important spending reductions are to come from a public sector pay freeze and reduced transfers to the regions. Yet, the problems beneath the debt crisis in Italy are mainly chronic ones that cannot be solved within a short period of time (Quirico 2010).

By 2011, Italy's public debt reached 120% of its GDP – the second largest in the Euro zone after Greece – and borrowing costs on sovereign government debt rose to record levels (CIA 2012). During the second half of 2011, the government passed a series of three austerity packages to balance its budget by 2013 and decrease its public debt burden (BBC 2012a; CIA 2012). These measures include a hike in the value-added tax, pension reforms, and cuts to public administration. In September 2011, Italy has its debt rating cut by Standard & Poor's, to A from A+ (BBC 2012a). The Euro zone crisis along with Italian austerity measures had reduced exports and domestic demand, slowing Italy's recovery (CIA 2012). Some critics commented that this crisis provides a vital opportunity to address Italy's long-term challenge—the chronic lack of growth (Quirico 2010).

Portugal

Portugal also encountered serious financial difficulties following Greece's bailout. The EU monetary integration brought forth a number of imbalances that contributed to Portugal's poor economic performance even before the 2008 financial crisis. For example, low interest rate and easy loans have allowed for the piling up of government debts. Since the mid-1990s, a fast rise in unit labor costs compared to core European countries also eroded Portugal's competitiveness (Euro Challenge 2012; OECD 2010). Over the years, Portugal's large current account deficits led to high levels of external debt, fuelled mainly by excessive private sector borrowing.

As the financial situation worsened, in May 2010 with support from the main opposition party, Portugal's government frontloaded the consolidation path and adjusted the deficit target from 9.3% of the GDP in 2009 to 7.3% in 2010 and then 4.6% in 2011 to foster foreign investor confidence (Euro Challenge 2012; Report Linker 2012). As time went by, the policies adopted apparently were not enough to achieve a 3% deficit target by 2013. As a result, Portugal implemented austerity measures, including a 5% public salary cut which went into effect on January 1, 2011 and a 2% increase in the value-added tax (CIA 2012). Without much option for stimulus measures due to its financial difficulties, the government was focusing instead on boosting exports and implementing labor market reforms to try to raise GDP growth (CIA 2011).

Unfortunately, some investors still expressed concern about the ability of Portuguese government to achieve these targets and cover its sovereign debt. In March 2011, Portugal's Prime Minister quit after the parliament rejected his latest austerity program. Thus, the political crisis heaped onto a debt crisis. After the government collapsed, Portugal was clobbered by a series of debt downgrades by the debt ratings agencies and rising bond yields (Reguly 2011). In early April 2011, the rate on Portuguese bonds reached almost 10%, close to triple the level paid by benchmark German bonds (Reguly 2011). Eventually, Portugal confirmed it would need a financial bailout from the European Union; in May 2011 a package of US$110 billion (€78 billion) was approved, following Greece and the Republic of Ireland (BBC 2012a; Minder 2011).

In the fourth quarter of 2011, the contraction in Portuguese economic activity had intensified. Exports registered slowdown while private consumption and residential investments had strong contractions. These effects were mainly due to the budgetary consolidation measures on household income prospects (Banco de Portugal 2011). For Portugal to reduce external imbalances, growth must be shifted from consumption to exports, which will require increased competitiveness through productivity gains in the medium term (Banco de Portugal 2011). To avoid jeopardizing the financing of the Portuguese economy, Bank of Portugal (Banco de Portugal) was authorized to permanently monitor the orderly and gradual deleveraging process of the financial system.

According to suggestions from the 2010 OECD Portugal Economic surveys, in the short term, Portugal could increase productivity by encouraging the entry of efficient firms and the exit of inefficient firms; in the medium term, further productivity gains could be achieved by continued climbing of the value-added chain in traditional exporting industries, such as textile and tourism; and in the long term, Portugal should expand industrial clusters, such as the "Competitiveness and Technology Hubs" and further develop cooperation programs between firms and the R&D sector with clear evaluation criteria (Euro Challenge 2012; OECD 2010).

In projecting the outlook of Portuguese economy in 2011–2013, Bank of Portugal suggested four internal matching measures to help achieve the projected goals. First, public administration reform should increase the efficiency in the provision of the goods and services and eliminate rents received by agents; second, a careful evaluation of all public expenditure programs is crucial for the structural reduction of the level of public expenditure; third, the implementation of the reforms in the labor market is crucial to boost job creation, such as increasing internal and external flexibility; and fourth, the reduction of debt levels of households and firms, in conjunction with the gradual and orderly deleveraging process of the banking sector (Banco de Portugal 2011).

Portugal needs to embark on a program of comprehensive reforms to raise its longer-term growth potential, correct its imbalances, and restart the convergence towards higher income-level countries in order to restore fiscal sustainability over the long run (OECD 2010).

Spain

Before the financial crisis, Spain's government had lower debt levels than Germany while spending less than it earned in taxes. Yet, the country experienced a housing bubble, thereby creating a surge in unemployment. Spanish banks' high exposure to the collapsed domestic construction and real estate market posed a continued risk for the sector and the country. As a result, Spain's banks increased their reliance on the European Central Bank's longer term financing operations for more than US$68 billion (€47 billion) in just one single month, from June and July 2009 (Hugh 2010).

Spain's large budget deficit and poor economic growth prospects had made it vulnerable to financial contagion from other highly-indebted Euro zone members. Spanish government had to oversee a restructuring of the savings bank sector in 2010, and provided around US$15 billion (€10.4 billion) in capital to various institutions. Particularly, the Bank of Spain was seeking to boost confidence in the financial sector by pressuring banks to come clean about their losses and consolidate into stronger groups.

In Spain, both households and corporations were excessively in debt. Corporate sector had around US$1.9 trillion (€1.3 trillion) debt (or 125% of GDP) in 2010 (Hugh 2010). In addition, wages inflated during the good years and pulled up by the overheating housing boom became uncompetitive as the wage growth outpaced that of other European countries (Harrington 2011). These developments significantly decreased investor confidence and raised the interest rates of Spanish loans. Furthermore, with Spain's rising debt, investors became more reluctant to invest and Spain's economy further declined (Harrington 2011).

Unfortunately, government efforts to boost the economy through stimulus spending, extended unemployment benefits, and loan guarantees did not prevent the unemployment rate from rising from a low of about 8% in 2007 to 20% in 2010 (Krugman 2010). The government budget deficit worsened from 3.8% of the GDP in 2008 to 9.2% in 2010, more than three times the euro-zone limit. Consequently, the Spanish government was forced into two austerity packages in 2010, both aimed at getting the budget deficit under control (see Appendix 1 for details). The target was to save US$7.2 billion (€5 billion) in 2010 and US$14.4 billion (€10 billion) in 2011 (Krugman 2010). To bring the deficit down from 11.2% of the GDP in 2009 to 3% by 2013, a series of tax measures were designed to cut the deficit by US$72 billion (€50 billion) over 4 years (Krugman 2010). For example, the government intended to raise money by increasing taxes on "the rich" – the wealth tax for all taxpayers with US$1.44 million (€1 million) or more in assets (Krugman 2010).

Spain also cut public wages by 5%, froze salaries and pensions in 2011, cut government benefits for new mothers, and raised the retirement age from 65 to 67 (Harrington 2011). These reforms also applied to regional governments, which hired almost half of all the public workers and controlled health spending. However, these policy changes still did not appear to be calming investors' worries. Moody's, a well-trusted credit rating agency, downgraded Spain's public bonds in March 2011, causing heavy stock market losses (Harrington 2011).

Spain was in a unique situation because of its real estate market and the actions of the unregulated cajas banking system reported in Chapter 2. When the real estate market collapsed, it brought the entire Spanish economy down with it. Because of the caja crisis, Spain has passed a new law requiring cajas to be more transparent in their lending practices and to reinforce their capital by September 2011 or face the prospect of partial nationalization (Harrington 2011). Some cajas have accepted the need for partial nationalization and will accept government money, while others are seeking outside investors and plan stock market flotation (Mallet 2011).

Although Spain still remained in a dire financial position, the Spanish government and the EU were working hard to ensure its recovery and stability (Harrington 2011). Ever since the global financial crisis, economic development in Spain has been closely watched. As a founding member and a representative of a larger entity in the Euro zone, if Spain were unable to pay back its sovereign bonds, analysts and investors feared that the entire Euro zone would be in jeopardy. The EU leaders also knew that it was imperative that Spain survive the economic crisis because of its economic power and size. Spain accounts for 11.5% of Euro zone GDP while Greece only accounts for around 2.5%.

However despite the fact that France and Germany tried very hard to prevent Spain from needing a bailout, still the Spanish government had to request European financing for the recapitalization of its failed banks after ratings agency Fitch hit Spain with a three-notch downgrade on its credit rating (Wolls and Dilorenzo 2012). On June 10, 2012, financial ministers in the Euro zone agreed to lend Spain up to US$125 billion (€100 billion) (RTE 2012; Wolls and Dilorenzo 2012). Since the deal imposed no conditions on the overall Spanish economy and no new austerity measures, the Spanish government emphasized that it was an aid not a rescue (RTE 2012).

Chapter 5
Future Perspectives and Policy Implications

By the middle of 2012, more than three years had passed since the outbreak of the 2008 global financial crisis. Although these four Southern European countries are still mired in financial trouble, it is about time to reflect upon what happened during the past few years and what can be learned from this crisis for future preventive actions.

The situation in these four countries continues to be clouded by significant uncertainties that investors face. Unlike the speedy recoveries of certain Asian economies, such as China and Hong Kong, there has been no drastic improvement to clearly indicate a return to norm. Yet, no matter whether a particular country experiences fast or slow recovery, lessons can be learned by going through the process of planning and implementing stimulus package and continuing national reform for sustainability in the future. In other words, the economic crisis provides an ideal opportunity to examine the soundness of a nation's economic system and the effectiveness of national governance. In this chapter, we first present the prospects of these four Southern European countries, and then describe their potential challenges in the future. What we describe hereafter is mainly based upon the OECD economic survey report for those countries, supplemented with other relevant reports. The prospects and challenges are meant to provide points of thought or discussion and are not an exhaustive list. This chapter finishes with some policy implications.

Prospects

Although these four Southern European countries have not totally recovered from the 2008 global financial crisis, there are still signs of promising prospects, which are briefly explained below.

C.Y.-Y. Lin et al., *National Intellectual Capital and the Financial Crisis in Greece, Italy, Portugal, and Spain*, SpringerBriefs in Economics 7,
DOI 10.1007/978-1-4614-5990-3_5, © The Author(s) 2013

Greece

In Greece, considerable advances have been made with structural reforms both in the public sector and in labor and product markets through this financial crisis. To establish a healthier financial system, the 2010 fiscal deficit was reduced mainly through across-the-board tax increases and cuts in civil service pay and pensions (OECD 2011a). Despite the short term costs, the reforms that have been planned and implemented will definitely benefit Greece for many years to come, as they will raise growth, living standards and equity. However, the prerequisite of success is that the burden and benefits of reform be broadly and fairly shared between the public and private sectors and among the citizens (OECD 2011a). In addition, the crucial success factor relies on thorough and impeccable implementation.

The reforms carried out by the Greek government over the past years are impressive. Cuts in the public deficit were unprecedented, the statistical system is being improved significantly, red tape and barriers to competition are being cut which boost investment and foreign directed investment (OECD 2011a). In addition, deep-rooted fiscal reforms have been launched to strengthen the management of the public finances and to revamp the pension system.

Although the Greek economy is still weak and the path to recovery is lengthy, its real GDP growth is estimated to stop the continuous drop at −3% in 2011 (Global Finance 2012), which indicates that the reforms may bear fruit. With such positive developments, authorities should continue this vigorous reform process so as to convince markets of their capacity to implement fundamental economic change. In addition, the Greek government should also convince the domestic audience of the need to pursue the reforms required to bring public finances to a sustainable path and to position the Greek economy for robust growth in the years ahead (OECD 2011a).

Italy

For Italy, in 2012 the economy has already passed through a deep recession triggered by the global crisis and seems set for a gradual recovery. In summer 2010, as soon as the first signs of economic recovery became apparent, Italy adopted a 3-year budget: a deficit correction of US$17.3 billion (€12 billion) in 2011 (0.8% of GDP) and almost US$36 billion (€25 billion) in each of the following two years (OECD 2011b). Since the current crisis provides an opportunity for defining a new model of development in Italy (Quirico 2010), continuous structural reforms, educational system reforms, and financial system reforms have been adopted by the Italian government.

The priority of Italian government is implementing structural reforms to increase growth potential, while maintaining a stable fiscal framework oriented towards consolidation. Such a policy can sustain confidence in Italian public finances in the face of the large stock of government debt. It in turn helps support the financial system, which is crucial for the recovery. The government is ambitious to reduce the overall deficit below 3% of GDP by 2012. Reform of the public administration has shown some initial success in improving transparency, but the deeper aspects of the plans to improve efficiency in public administration must be continuously pursued (OECD 2011b). The use of various audit mechanisms, such as regulatory impact analysis and public spending reviews, should help bring some positive effects.

In reforming the educational system, a new quality assurance step is to develop reliable indicators of performance to aid both students and universities in achieving better performance. Further actions to facilitate university-enterprise research collaboration, especially through allowing contracts that permit an efficient share of costs and benefits between individual researchers, their institutions and the private sector, would help to improve university finances and would boost Italy's weak showing on many measures of R&D and innovation performance (OECD 2011b). Emphases on innovation in environmentally-friendly technologies are also encouraged to increase the economy's ability to deliver "green growth" (Gurria 2011b).

As for financial reform, measures taken by the Bank of Italy were in line with those defined at the international level; including the recommendations of the Financial Stability Forum (FSF). In addition, banks support measures adopted by the Italian government, including the emergency recapitalization scheme, the funding guarantee scheme, the swap facility, and the full state underwriting of the deposit insurance schemes; all these have substantially reduced systemic risks as well as liquidity and funding strains, thus set the legal framework for other actions as needed (IMF 2008). On the way to Italy's recovery, the above described focuses and policies may help Italy regain its past strength.

Portugal

In 2010, Portugal increased the VAT rate and reduced tax expenditures in direct taxes. These were positive approaches for its economic recovery. In addition, Portugal has launched a number of ambitious measures to enhance its transport infrastructure, such as a high-speed railway and port platforms, partly aimed at enhancing the connections with Spain.

Furthermore, over the last three years, important measures have been taken to strengthen teachers' skills and improve the quality of teaching. In particular, Portugal has introduced teacher performance evaluation at the primary and secondary level (OECD 2010), which somewhat guarantees the quality of teaching. Although positive steps have been taken to correct the main chronic problems in Portugal, success relies upon the lasting efforts and reforms, not only to recover but also to advance for future sustainability.

R&D has appeared to be the strength of Portugal compared with the other three countries, as reported in Chapter 3. Since innovation and renewal capability become the key drivers for future economic growth, more and more countries are increasing investments in nurturing new innovation and new industries. The continuous investment in R&D and renewal capital improvement of Portugal are positive signs for its future development, even though it has not yet totally recovered from the financial crisis in 2012.

Spain

Spain still possess good fundamentals required for the financial recovery, as it has enjoyed democratic rule and good economic performance in business, trade and investment freedom with an overall entrepreneurial environment supporting private-sector development in the past (Heritage 2011). Even as recently as 2007, Spain was seen as a model economy while being ranked one of Europe's richest economies. Although the situation worsened over the past few years, resuming to its previous status should be relatively easier for a nation with this kind of success record.

From this debt crisis experience, the Spanish economy will now inevitably have to move from being a construction and tourism driven one to a more broadly-based profile, as the continuing demise in construction is a logical and inevitable part of the ongoing adjustment. Spanish government also knows that what Spain badly needs is a trade surplus, which helps reduce the banks' heavy reliance on a massive liquidity injection from the European Central Bank, as it is very likely that these will be withdrawn someday. Following the drop in the value of the Euro, exports have revived somewhat (Hugh 2010). This may help Spain to regain its past strength faster.

Challenges

Although governments of the four Southern European countries have strategized to cope with the debt crisis for future sustainability, a lot of challenges remain to be tackled. This section briefly introduces the issues observed that need special attention for each of the countries.

Greece

In Greece, the government used to take advantage of greater access to cheap credit to pay for government spending and offset low tax revenue. The government also borrowed to pay for imports from abroad that were not offset by exports overseas. Government budget and trade deficits ballooned during the 2000s and the borrowed funds were not channeled into productive investments that would generate future growth, increase the competitiveness of the economy, and create new resources with which to repay the debt. Instead, the inflows of capital were used to fund current consumption that did not yield streams of revenue with which to repay the debt (Nelson et al. 2011).

Despite the supports from European Union, European Central Bank, IMF and Greek government reforms, persistent market skepticism concerning the capacity of Greece to restore sustainable public finances and to renew economic growth remains. Therefore, Greek government needs to implement essential policies to cope with the challenges stipulated below to restore market confidence and to rebuild its stale economy.

The most recent OECD Greek Report (OECD 2011a) warned the Greek government that its reform program may face political risks, as austerity measures will definitely impact some vested interest groups. In addition, the condition of a gloomy world economy outlook, rising commodity prices and unsustainable debt burdens in many OECD countries also poses risks as it slows down economic recovery.

However, the most difficult challenge for the Greek government is to reform itself – the public sector system (OECD 2011a). For instance, the public payroll has grown rapidly over the past decade. Low and middle-rank civil servants have higher wages than similar private sector employees although they work on average fewer hours and have greater job security. Data published by the OECD show that the Greek government spending is close to half the country's GDP every single year between 1995 and 2008. Furthermore, a significant portion of the spending went not to stimulate the economy out of recession, but to pay for social welfare programs with questionable returns on economic growth and vitality. OECD data points to another example of Greek government liberal social spending – workforce retirement. The retirement age in Greece is 58 years, which is the lowest among all Euro zone and OECD countries compared to 65 in Germany, UK, Italy and Spain (OECD 2011a). In other words, the Greek government had a huge entitlement program, a stagnant economy and dysfunctional tax collection system. The government also looked to sovereign debt to generate liquidity to pay for social entitlements and other fiscal obligations. Persistent over-borrowing and over-spending aggravated Greek government deficit, which climbed higher every year.

Now with external supports and a second bailout, the additional time that Greece has must be used in the best possible way. In summary, the challenges it faces include:

1. Stepping up the pace of its reform effort adopting a firm and unified stance behind the reform program to enhance credibility, and publishing widely reform results under the scrutiny of its social partners.
2. Setting the scope timing, and quality of economic statistics to enhance the diagnostic tools required for monitoring its economy.
3. Improving its budget monitoring system and establishing a better fiscal framework.
4. Greece can consider putting its fiscal rule in its Constitution, as Switzerland and Germany have done, to show its commitment to sound fiscal policy.
5. Establishing clear responsibilities in ministries for monitoring reform implementation on the basis of objective indicators and a rigorous legislation to monitor its government effectiveness.
6. The Bank of Greece should maintain its attentive supervision of individual credit institutions.

7. Tackling tax evasion decisively as currently personal income tax revenues are around 5% of the GDP below the euro area average.
8. Overhauling the public remuneration system and restructuring state-owned enterprises including significant wage cuts, controlling spending, enhancing the quality of public services, and increasing the number of hours worked.
9. Swiftly reducing cumbersome business regulations that have blunted entrepreneurship job creation and investment to create a business-friendly Greece.
10. Getting people back to work with active labor market policies and by reforming labor market outcomes.
11. Reforming the health sector to better control medical prescriptions hospital expenditures, public procurement and the introduction of a global hospital budget system.

The potential impediment of the above challenges is market confidence, as it has not been regained in early 2012. Since markets and the domestic audience have doubted the ability of Greek government's reforms to deliver growth and sustainable fiscal consolidation, the key success factor relies on "effective implementation" to actually show the planned results for market confidence.

Italy

In striving to regain competitiveness, Italy faces a series of challenges elaborated below (CIA 2012; IMF 2008; Quirico 2010):

1. Italy's main economic strength was its large base of small and medium enterprises (SMEs), which suffered from high labor costs. In order to cut costs and gain bigger profits, Italian SMEs began to move their manufacturing plants to East Europe and Asia, which led to a decrease in jobs, an increase in unemployment and greater pressure on public spending in Italy.
2. Italy has a considerable economic gap between its northern and the southern areas. The annual average income of people in the south was only 75% of that in the north in 2007. In order to boost economic development, the Italian government kept investing in the south, which, to their disappointment, had little effect.
3. Italian government needs to appropriately improve management of public assets. In addition, maintaining fiscal discipline requires a transparent, formula-based system of equalization transfers, a robust and independent regime for monitoring fiscal targets, harmonized accounting, and greater civil service mobility.
4. Italy has not been strong in high-tech fields such as bio, electronic, and information, making it more difficult to develop its high-value added industries. Therefore, it is less competitive in the knowledge economy.
5. In Italy, there is little connection between research and production, which hampers the potentiality to move towards a high-tech economy. Normally, Italian firms have

to work with foreign patents and rely on foreign research activities. As a result, such a context does not allow advanced industries to take roots in Italy.

6. There has been no innovation in government industrial policy. Job losses have obviously been the most significant social impact of the crisis. In this type of situation, the unemployment rate will remain high and may develop social unrest.
7. Although the Italian government has struggled to limit government spending through structural reforms, such as reducing graft and overhauling costly entitlement programs, it still has high public debt and a high fiscal deficit that far exceeds EU standards. Extra efforts are needed in order to achieve the planned goals.
8. Italian government needs to deal with the persistent problems of illegal immigration, organized crime, corruption, high unemployment and sluggish economic growth. Italy also has a sizable underground economy, which by some estimates accounts for as much as 15% of its GDP.
9. In the longer term, Italy's low fertility rate and quota-driven immigration policies will increasingly strain its economy.
10. In general, fiscal policy, structural policy, university sector, and environmental policies are the four major points that need to be reformed for Italy's future sustainability.

In addition to the above challenges, Italy's education system needs both to instill open and forward-looking values in young people and to prepare them for the future job market, since without a proper value system the challenges posited above cannot be overcome for the overall benefit of Italy (Quirico 2010).

Portugal

The most problematic factors for doing business in Portugal are inefficient government bureaucracy and restrictive labor regulations (Reguly 2011). Another chronic problem is its tax system, characterized by narrow tax bases and higher tax rates, with tax collection losses and stagnant growth; in addition, property taxation in Portugal remains below the OECD average. Apparently, Portugal will have to go through a lengthy path to recovery and needs to face the following challenges (OECD 2010; Reguly 2011):

1. Reducing its deficit, tax expenditures, and the scope of public enterprises, while increasing VAT rate.
2. Increasing efficiency in state-owned enterprises by expanding performance monitoring mechanisms and concentrating on the efficiency of firms, labor productivity and labor flexibility.
3. Improving flexibility in working-time regulations, such as job subsidies and short-time work to reduce cyclical employment losses and to promote productivity growth.

4. Reforming its tax system to move towards a more growth-friendly and equitable tax system such as VAT and property taxes, and to become business-friendly and thereby induce more investment and increase job opportunities.
5. Reforming the education system by evaluating teaching quality, reducing the amount of early school dropouts, and up-skill the labor force.
6. Embarking on a program of comprehensive reforms to raise its longer term growth potential, correct its imbalances, and restart the EU convergence process.
7. Enhancing transport infrastructure, high-speed railway and port platforms.
8. Reducing greenhouse gas emissions from the transportation sector.

Spain

Spain is fully aware that the soundness of its economy is very important to the Euro zone recovery. With a widening deficit, the government needs to deal with the dilemma of cutting spending and raising taxes, and risk deepening the slump due to the fact that austerity may produce more austerity (Samuelson 2012). Yet, the absence of austerity may produce a crisis of confidence in the international financial markets, thus Spain will not be able to borrow more capital at reasonable interest rates. Challenges Spain has to face include the following issues:

1. The greatest challenge Spain faces is the aftermath of its housing construction bubble burst. A key player behind this overheated housing tragedy is the liberal lending offered by the unregulated private banks, or "cajas." Therefore, Spain must find ways to handle the cajas and their real estate debt. Although the Spanish government is strongly "encouraging" (by threatening nationalization) the cajas to recapitalize on their own by attracting private investor capital, legal enforcement may be needed.
2. After the housing bubble, the Spanish economy has to move from being construction and tourism driven to having a more broadly based profile with growth potential.
3. Government initiated strategy to build a future national growth engine needs to be formulated as soon as possible for resource preparation, including manpower.
4. Various ministries and institutes, as well as the private sector, have to fairly share the responsibility for accomplishing the planned goals of spending cuts, privatizing industries, and boosting competitiveness through labor market reforms.
5. Spanish government needs to effectively communicate and integrate joint efforts to deliver the results in order to regain market confidence.
6. Another future challenge is its low birthrate, which is lower than the EU average at 1.48 births per woman in 2012 (CIA 2012), and well below the "replacement rate" of 2.1.

Policy Implications

The main cause for the debt crisis of the four Southern European countries is the breakdown of their financial surveillance mechanism. Maintaining fiscal discipline will require a transparent financial system, harmonized accounting, a robust and independent regime for monitoring fiscal targets, sound public assets management (IMF 2008), and effective national governance. These characteristics manifest that intangibles – people, systems and governance – are the key success factors. National intellectual capital (NIC) represents national intangible assets to a certain degree. Research results reported in this booklet provide valuable information for the policy makers to look back on the national intangibles before the crisis, observe their codevelopment with the GDP during and after the crisis, and then draw lessons from it.

By investigating economic development and NIC development from 2005–2010, we have been able to connect the co-evolution of tangibles and intangibles for the four Southern European countries. Implications that can be drawn from our research findings are as follows:

First, national intellectual capital development goes together with the economic development and should be regarded as an enhancer of economic growth. The co-development of NIC and GDP with a strong positive correlation (Fig. 3.7) indicates the higher the national intellectual capital the higher the GDP. This means that intangible intellectual capital development accompanies GDP growth. Figures 3.8, 3.9, 3.10, 3.11, and 3.12 further show that NIC development reflects the economic difficulties of these four countries during the financial crisis. Figures 3.17, 3.19, 3.21, and 3.23 highlight the fact that Greece and Portugal lost their international competitiveness from market capital and process capital after the financial crisis. The early signs of decline for all the four countries (starting in 2007) were also uncovered in Figs. 3.15 and 3.16. Therefore, attending to the development of NIC should help policy makers gain different perspectives on economic growth and identify some early signs for timely national strategies.

Second, more effective national and NIC governance systems need to be established. At the onset of this global financial crisis, these four countries were not seriously impacted. It was the trade decreases, the tightening of international money market, and the decade long over-lending and over-spending that brought their financial systems down, especially in Greece and Portugal. A main problem is the failure of the surveillance system. Figure 2.2 shows that the general government debt of Greece, Italy, and Portugal already exceeded the 60% of GDP rule as early as 2005, and for Greece it has been a chronic, decade-long problem. If the governments would have attended to such financial problems and seriously resolved them, their economies would have been more resilient to this financial crisis. In addition, setting up a NIC governance system by continuously monitoring NIC development should help detect early warning signs of potential risk.

Third, national statistics need to be carefully studied, refined for NIC purposes, effectively interpreted, and strategically utilized for NIC policy. In hindsight, many statistics before the financial crisis have revealed important messages early enough for preventive actions. Taking Greece as an example, other than its decade-long high government debt, its public spending of almost 50% of the GDP, doubling the wages of civil servants, and a retirement age of 58 (compared to the general 65 of other countries) are all signs of un-sustainability. The effect of salary raise in the public sector means salary raise in general, decreasing national wage competitiveness, driving companies to relocate to low wage countries, reducing job opportunities, and then hiking up unemployment rate.

Another example is the housing loan in Spain. As reported previously, in order to promote integration, loans for more than 100% of a house's price were common in Spain. This stimulates housing demands and results in increased builders, more low-skill immigrants, more houses being built and sold, more unsustainable loans, and a drop in overall national productivity due to the increasing of a low-skill workforce. When the financial crisis hit, the housing bubble burst and the financial system almost crashed. This example shows that every policy change needs to be carefully calculated to account for its side effects, short-term and long-term aftereffects, international comparability, and sustainability. Particularly, national policies should not fall prey to politics for winning election votes. Interpreting government statistics together with NIC development may unveil important directions for mapping the development of national tangible and intangible assets. For example, Greece's capital availability as a GDP growth impeding factor has significant meaning for tracing the soundness of national financial system as well as the changes of market capital.

Fourth, educating public officials on effectively managing public assets for the benefits of the whole country is essential. The debt crisis of each one of these four Southern European countries reveal a general lack of vision and determined execution ability. Other than the absence of EU's oversight to prevent the crisis from happening, high ranking government officials should take responsibility as well. Although running a nation is much more sophisticated than leading a company, the principles of leadership are similar. In the business world, effective monitoring systems are crucial for business success. The chief executive officer, general manager and chief financial officer need to read various statistics and reports almost daily in order to detect potential problems for timely coping strategies. Publicly available statistics presented in this booklet have disclosed problematic numbers before the financial crisis, as the two cases described in implication number three above so attest. It is surprising that public officials in charge let obvious problems develop to a detrimental level and then burst. Therefore, the development of visionary public officers with expert domain knowledge is critical to effectively run and manage public assets for the benefit of the whole country.

Fifth, national pride and trust should be cultivated within the citizens for community and social capital platform evolution. The World Economic Forum defined these four countries as Stage 3 innovation-driven countries, yet their Global Competitive Index is low among the Stage 3 countries (see Table 1.1), their national intellectual capital ranges from 26 to 31 out of 48 countries, and the level of their renewal capital is grouped together with the developing countries rather than the developed countries (Fig. 3.12). Being advanced countries based on ancient Greek civilization (Greece), the Roman Empire (Italy), and ocean super powers (Portugal and Spain), their contemporary economic situations have become uncompetitive, to the extent that Greece and Portugal need financial bailouts, and Spain needs financial aid. This global financial crisis provides a good opportunity for these countries to contemplate ways to regain their past glory. Most importantly, national pride and trust need to be cultivated and a social capital platform established as soon as possible.

Sixth, utilizing and in-sourcing human capital to revitalize and energize economies may be a plausible measure. After the financial crisis, these four countries need to continue some of the austerity measures to bring their economies back to health. As a result, their leverage becomes limited. Our research reveals that they still have relatively good human capital. From Figs. 3.29, 3.30, 3.31, and 3.32, we see that their human capital is in the middle capability level (except Greece) among the 48 countries. Although Greece's human capital is low, it did not regress during the financial crisis (Fig. 3.9), representing capital stability. Therefore, these four countries may utilize their relatively competitive human capital to pull up their stale economies. Losing wage competitiveness to emerging East European countries is a major reason for their faltering economies. In addition, providing more extensive employee training to increase human resources added value, in-sourcing human capital by importing skilled workers can also be considered. Therefore, policy makers of these four countries are encouraged to pay more attention to upgrading their current workforce and establishing a refined immigration policy to better in-source human capital for the revitalization of their economies.

Seventh, governments need to aggressively launch structural reforms for a more business-friendly and value-creating internal, as well as external, environment to boost future economic development and well-being. Our 3D NIC development trajectory unveils impeding factors to GDP growth for these four countries, summarized in Table 3.3. Frequency count indicates that government-related impeding factors has the largest total count of 20, including convenience of establishing new firms, corporate tax encouragement, fair business competition environment, government efficiency, corporate tax encouragement, and transparency of government policy. That is, the above government-related issues have hampered these nations from achieving the targeted GDP. The finding provides clear guidelines for the governments to effectively implement coping measures to retain companies inside their national borders, provide more job opportunities, and facilitate a speedier recovery and the well-being of the whole society. By improving the business environment, the national image of these four countries may also be enhanced, as "image of country" has been identified as a critical factor leading to the expected GDP (Table 3.4).

Eighth, economies with faster recovery from the financial crisis have higher national intellectual capital than those with slower recovery. To observe the relationship between NIC and the financial crisis, Fig. 5.1, containing two graphs, is presented to compare countries with different levels of recovery after the financial crisis.

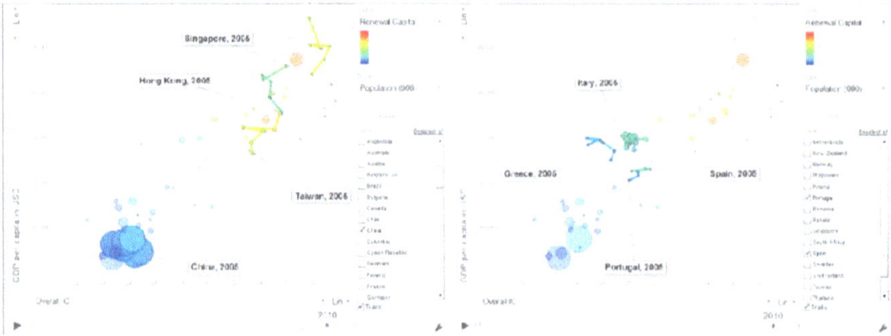

Fig. 5.1 Comparison of NIC and GDP co-development of the four Southern European countries and the Greater China economies

The right graph (Fig. 3.8) is for the four countries reported in this volume; the left graph is the same map using the same data set for economies in Greater China (China, Hong Kong, Singapore, and Taiwan). The four Southern European countries are still in financial trouble in 2012, yet economies in Greater China experienced speedy recovery starting from the last quarter of 2009. It is clear that Hong Kong, Singapore, and Taiwan have much higher NIC than the four Southern European countries. In addition, the scale of development each year for the former is larger than that of the latter. This is additional evidence that countries with better NIC have stronger recovering capability than those with weaker NIC. Therefore, investing in developing national intellectual capital should generate good prospects for national development.

Concluding Remark and Emerging Insights

In the aftermath of the 2008 global financial crisis, speed of recovery varies in different parts of the world. Although the situation in the EU is still unstable as of mid-2012, in the U.S. recovery began to show in the third quarter of 2009 (Kehoe 2010), and countries in East Asia have experienced faster recovery than the other regions. According to the World Economic Forum 2011–2012 report, at the beginning of 2011, worldwide recovery appeared fairly certain, with economic growth for 2011 and 2012 projected by the International Monetary Fund (IMF) at 4.3% and 4.5%, respectively.

Economic developments in the past few years reinforce the observation that economic growth is unequally distributed, which leads to the shift of economic activity from the west to the east. Although significant steps have already been taken by EU authorities to tighten financial regulations and rebalance economies, it is a difficult and prolonged process with such a large scale of imbalances. According to research, reducing public debt to pre-crisis levels will constrain government expenditures for at least a decade (Schwab 2011) and on average, it takes four and a half years to get back to the same per capita GDP where a country started out and about the same amount of time for unemployment to stop rising (Rogoff 2011).

As a recap, to counter worsening economic conditions worldwide, a total of US$2 trillion in stimulus packages, amounting to approximately 3% of the world gross domestic product, has been allocated (some of the packages include measures that extend into subsequent years), an amount that exceeds the call by the International Monetary Fund (IMF) for fiscal stimulus totaling 2% of global GDP. IMF also estimated that as of January 2009, the U.S. fiscal stimulus packages as a percentage of the GDP in 2009 would amount to 1.9%, for the Euro area 0.9%, for Japan 1.4%, for Asia excluding Japan 1.5%, and for the rest of the G-20 countries 1.1% (Nanto 2009).

This global financial crisis came with unexpected speed. Although there were early signs of problems in public statistics, they were largely neglected by countries seriously in trouble, such as the four Southern European countries investigated in this booklet. When the financial crisis hit, some deficit countries faced the combined problems of a sharp contraction in private demand, an impaired financial system and weak public finances, which overwhelmed the governments in deep trouble. Looking back reveals that the debt crisis in these four countries could have been prevented; if the governments had followed the rules regulated by the European Union and fiscal information were transparent.

In the wake of the crisis, a new surveillance mechanism is in place to initiate necessary procedure and to propose economic sanctions on a member state whose structural imbalances are not corrected (IMF 2008). Important legislative changes are also underway at EU level to improve the surveillance of imbalances and to help ensure that the necessary corrective action is undertaken where risks emerge (Barnes 2010). Alongside with planned structural reforms, macroeconomic, financial and fiscal management would be strengthened in an integrated way in order to reap the best result. Some countries have also been experimenting with creating an independent fiscal council to monitor and assess official fiscal projections. Furthermore, structural reform efforts were made to use more output-based performance measures and to choose the measures with care to avoid distorting incentives.

Generally, countries impaired by financial crisis tend to develop a protection mentality, which reduces international interactions. To facilitate a faster recovery, G-20 members have continued to honor their pledge not to retreat into investment protectionism. As a result, most of their policy measures continue to move towards eliminating restrictions on foreign investment and improving transparency for investors, which set a good example for countries worldwide to follow.

In terms of future NIC development, our other study on the top ten NIC ranking countries reveals that the interplay among the four capitals – human capital, market capital, process capital, and renewal capital – resulted in the best synergetic prospect for GDP growth. When their weaker capital (process capital) interacts with stronger capital (market capital), both effects on the GDP are enhanced (Lin 2011). This finding prompts our suggestion for the four countries to enliven their stale economies by capitalizing on their relatively competitive human capital.

Overall, NIC intelligence suggests that in an era when the intangible asset has become a key competitive advantage, investing in national intellectual capital development is investing in future economic development and national well-being. To prevent a similar financial crisis from happening again, what Rogoff (2011) has suggested is on the mark: "I think public-finance experts need to methodically go through the system and strip debt subsidies out. I would start with changing our corporate-tax law and any overt incentives that favor debt." In addition, Rogoff (2011) continued, "The most damaging effects of financial crisis come not from loss of wealth, but rather from those who are unprepared for such losses and panic in response." Presumably, national leaders should have learned some valuable lessons from this financial crisis. Transparent information from multiple perspectives, including the NIC intelligence reported in this volume, will enable decision makers to have better sensibility in preventing a similar situation from happening again and to not panic when in deep trouble.

Appendices

C.Y.-Y. Lin et al., *National Intellectual Capital and the Financial Crisis in Greece, Italy, Portugal, and Spain*, SpringerBriefs in Economics 7,
DOI 10.1007/978-1-4614-5990-3,

Appendix 1 Summary of the main stimulus packages of the four Southern European countries

Country	Date	Amount	Item	Sponsor and remarks
Greece	May 2010	#1	Bailout	US$115bn (€80bn) by EU, European Central Bank (ECB)
		US$158bn (€110bn)		US$43bn (€30bn) by IMF
	May 2010–June 2011	#1 US$65bn (€45bn)	Greek bonds	ECB purchased Greek bonds (estimation of market analysts)
	July 2011	#1	2nd bailout	EU, ECB
	March 2012	US$157bn (€109bn)		
		#14		(proposed)
		US$172bn (€130bn)		17 finance ministers of EU members formally approved
	Oct. 27, 2011	#2	Write-off	Private banks holding Greek debt accept a write-off of 50 % of their return
	Implemented in 2011	#1 US$40bn (€28bn, about 12 % of GDP), including US$9bn (€6.5bn, about 2.9 % of GDP) spending cuts and revenue measures	A consolidation program through 2015	Greece's medium-term fiscal strategy (MTFS)
	By 2015	#1 Raising US$72bn (€50bn), including US$22bn (€15bn) by 2013	Privatization and public real estate development program	An ambitious privatization and public real estate development program

Italy	Nov. 2008	#3 US$3.8bn (€3bn)	Anti-crisis plan, aids to low-income households	€2.3bn—tax rebates for companies €0.7 m—tax rebates for households
	Feb. 2009	#4 #5 #6 US$103bn (€80bn), including the following items:	Stimulus package	Tax breaks for companies and poor households
		#5 US$3bn (€2.4bn)	Tax deductions	
		#6	3-year program	Up to US$1,950 for trading in an old car for a less polluting one
		US$2.56bn (€2bn)		20 % tax deductions for purchases of appliances and furniture
		#6 US$1bn		For building a bridge and increasing welfare aid
		#5		Delay payment of VAT, mortgage rate lowers down to 4 %
		#4 US$25.6bn (€20bn)		State helps Italian banks to ensure continuous support for lending
		#5 US$23.1bn (€17.8bn)		Invested in infrastructures
Portugal	December 13, 2008	#1 #7 US$2.92bn (€2.18bn) (1.25 % GDP), among which €1.3bn from government budget	Investment and Employment Initiative	Incentives for SMEs and exports, support for employment and for renewable energies
	April 2011	#8 US$115.56bn (€80bn)	EU bailout	EU, IMF
	May 2011	#9 #15 US$110bn (€78bn)		Formally approved. EFSM, EFSF, and the IMF, each contributes 26bn euro

(continued)

Appendix 1 (continued)

Country	Date	Amount	Item	Sponsor and remarks
Spain	2008	#10 US$66 bn (€50 bn) in stimulus, including the following two items:	Government stimulus package	8 billion euro has been distributed directly to local authorities for public works projects meant to create 200,000 jobs
	August 2008	US$29.4bn (€20bn)		Aids to SMEs and construction of social housing
	Nov. 2008	US$13.8bn (€10.9bn)		Hire unemployed Expenditures in R&D Aids to auto industry
	June–July 2009	#11 Increase from US$66.5bn (€47bn) to US$111.8bn (€79bn)	European Central Bank loan	Spain's banks increased their reliance on the ECB's longer-term financing operations
	May 2010	#12	Consolidation program	Freeze on state pensions, cut public sector salaries, up to 15 % for the top paid but averaging 5 % overall, abolish the "baby check," a US$3600 (€2500) payment made to all mothers of new babies
	March 2011	#13 FROB disbursed about US$16.8bn (€12bn)	Fund for Orderly Bank Restructuring (FROB) loan	Provide funds to manage bank restructuring, banks must follow the rules laid down by the government, and banks eventually must repay the FROB loans
	June 10, 2012	#1 to #17 US$125 billion (€100 billion)	Financial aid	EU, approved by 17 financial ministers in EU zone

Sources:
1. Nelson RM, Belkin P, Mix DE (2011) Greece's debt crisis: overview, policy responses, and implications. Congressional Research Service Report for Congress. Retrieved form http://www.fas.org/sgp/crs/row/R41167.pdf
2. Rowley A, Plowright M (November 16, 2011) Banks throw down gauntlet on Greek debt deal. Emerging markets. Retrieved from http://www.emergingmarkets.org/Article/2935964/Banks-throw-down-gauntlet-on-Greek-debt-deal.html
3. OECD (n.d.) Italy fiscal stimulus package and strategies to reduce fiscal deficit. Retrieved from http://www.oecd.org/dataoecd/42/30/47741933.pdf

4. Economy Watch (June 30, 2010a) Italy economic stimulus package. Retrieved from http://www.economywatch.com/economic-stimulus-package/italy.html
5. Dalje.com. (November 28, 2008) Italy approves stimulus plan, analysts skeptical. Retrieved from http://dalje.com/en-economy/italy-approves-stimulus-plan-analysts-sceptical/208950
6. Nanto DK (2009) Global financial crisis: analysis and policy implications. Diane Publishing
7. Economy Watch (June 30, 2010b) Portugal economic stimulus package. Retrieved from http://www.economywatch.com/economic-stimulus-package/portugal.html
8. Hall J, Rowley E (April 7, 2011) Portugal forced to seek €80bn EU bail-out. The telegraph. Retrieved from http://www.telegraph.co.uk/finance/economics/8433980/Portugal-forced-to-seek-80bn-EU-bail-out.html
9. BBC (March 30, 2012) EU austerity drive country by country. Retrieved from http://www.bbc.co.uk/news/10162176
10. Startfor (April 28, 2009) The financial crisis in Spain. Retrieved from http://www.stratfor.com/analysis/20090428_financial_crisis_spain?page=20
11. Hugh E (May 8, 2010) Spain emerges from recession? A fistful of Euros. Retrieved from http://fistfulofeuros.net/afoe/spain-emerges-from-recession/
12. Krugman P (February 5, 2010) The Spanish tragedy. New York Times. Retrieved from http://krugman.blogs.nytimes.com/2010/02/05/the-spanish-tragedy/
13. Harrington C (April, 2011) The Spanish financial crisis. The University of Iowa Center for International Finance and Development. Retrieved from http://blogs.law.uiowa.edu/ebook/content/spanish-financial-crisis
14. Castle S (March 14, 2012) With details settled, a 2nd Greek bailout is formally approved. The New York Times. Retrieved from http://www.nytimes.com/2012/03/15/business/global/greece-gets-formal-approval-for-second-bailout.html
15. BBC (May 16, 2011) Portugal's 78bn euro bail-out is formally approved, BBC News. Retrieved from http://www.bbc.co.uk/news/business-13408497
16. RTE (June 10, 2012) Eurozone agrees deal to lend Spain up to €100bn as Economy Minister Luis de Guindos confirms request for financial assistance, RTE News. Retrieved from http://www.rte.ie/news/2012/0609/spain-bailout.html
17. Wolls D, Dilorenzo S (June 10, 2012) Europe bailout of Spain could cost $125 billion. Yahoo News. Retrieved from http://news.yahoo.com/europe-bailout-spain-could-cost-125-billion-190614499--finance.html

Appendix 2 Important meetings held by world leaders to address the 2008 global financial crisis

Date	Place	Participants
June 18–19, 2012	Los Cabos, Mexico	G20
June 17, 2012	Los Cabos, Mexico	G7 finance ministers
March 1–2, 2012	Brussels	European Council
January 30, 2012	Brussels	Informal meeting of members of the European Council
December 8–9, 2011	Brussels	European Council
November 12–13, 2011	Honolulu, USA	19th APEC Economic Leaders' Meeting
November 3–4, 2011	Cannes, France	G20
October 23, 2011	Brussels	European Council
September 10, 2011	Marseille	G8 Deauville Partnership finance ministers
September 9–10, 2011	Marseille	G7 finance ministers and central bank governors
July 21, 2011	Brussels	Meeting of heads of state or government of the Euro area
June 23–24, 2011	Brussels	European Council
March 24–25, 2011	Brussels	European Council
March 17, 2011	Japan	G7 finance ministers telephone conference
March 11, 2011	Brussels	Extraordinary European Council
February 4, 2011	Brussels	European Council
December 16–17, 2010	Brussels	European Council
November 21, 2010	Ireland	G7 finance ministers
November 13–14, 2010	Yokohama, Japan	18th APEC Economic Leaders' Meeting
November 11–12, 2010	Seoul, Korea	G20
October 28–29, 2010	Brussels	European Council
October 8, 2010	Washington DC	G7 finance ministers and central bank governors
September 16, 2010	Brussels	European Council
June 26–27, 2010	Toronto, Canada	G20
June 17, 2010	Brussels	European Council
March 25–26, 2010	Brussels	European Council
February 11, 2010	Brussels	Informal meeting of heads of state or government of the European Union
February 5–6, 2010	Iqaluit, Nunavut, Canada	G7 finance ministers and central bank governors
December 10–11, 2009	Brussels	European Council
November 19, 2009	Brussels	Informal meeting of heads of state or government of the European Union
November 14–15, 2009	Singapore	17th APEC Economic Leaders' Meeting
October 29–30, 2009	Brussels	European Council
October 3, 2009	Istanbul, Turkey	G7 finance ministers and central bank governors
September 24–25, 2009	Pittsburgh, USA	G20

(continued)

Appendix 2 (continued)

Date	Place	Participants
September 17, 2009	Brussels	Informal meeting of heads of state or government of the European Union
June 18–19, 2009	Brussels	European Council
June 12–13, 2009	Lecce, Italy	G8 finance ministers
April 24, 2009	Washington, DC	G7 finance ministers and central bank governors
April 2, 2009	G20 London Summit United Kingdom	G20
March 19–20, 2009	Brussels	European Council
February 2009	Rome	G7 finance ministers and central bank governors
December 11–12, 2008	Brussels	European Council
November 2008	G20 APEC Peru	G20
November 22–23, 2008	Lima, Peru	16th APEC Economic Leaders' Meeting
November 14–15, 2008	Washington, DC, USA	G7
October 15–16, 2008	Brussels	European Council
October 10, 2008	Washington, DC	G7 finance ministers
September 1, 2008	Brussels	Extraordinary European Council
July 7–9, 2008	Toyako, Japan	34th G8 Summit
June 19–20, 2008	Brussels	European Council
June 13–14, 2008	Osaka	G8 Finance Ministers' Meeting
March 13–14, 2008	Brussels	European Council
February 9, 2008	Tokyo	G7 Finance Ministers' and Central Bank Governors' Meeting

For further information:
http://g20mexico.org/en/g20/previous-leaders-summits
http://www.g8.utoronto.ca/finance/index.htm
http://video.consilium.europa.eu/ecofin/
http://www.european-council.europa.eu/council-meetings/conclusions
http://www.apec.org/

Appendix 3 Indicators in each type of capital

Human capital index	Market capital index
1. Skilled labor*	1. Corporate tax*
2. Employee training*	2. Cross-border venture*
3. Literacy rate	3. Openness of culture*
4. Higher education enrollment	4. Globalization*
5. Pupil-teacher ratio	5. Transparency*
6. Internet subscribers	6. Image of country*
7. Public expenditure on education	7. Exports of goods

(continued)

Appendix 3 (continued)

Process capital index	Renewal capital index
1. Business competition environment*	1. Business R&D spending
2. Government efficiency*	2. Basic research*
3. Intellectual property rights protection*	3. R&D spending/GDP
4. Capital availability*	4. R&D researchers
5. Computers in use per capita	5. Cooperation between universities and enterprises*
6. Convenience of establishing new firms*	6. Scientific articles
7. Mobile phone subscribers	7. Patents per capita (USPTO + EPO)

Remarks:
Financial capital is the logarithm of GDP per capita adjusted by purchasing power parity.
Indicators marked with an asterisk (*) are rated qualitatively using a scale of 1–10.
Indicators with objective numbers are transformed to 1–10 scores, by calculating the ratio to the highest number in that indicator and then multiply 10.
Readers can refer to the book *National Intellectual Capital: A Comparison of 40 Countries* for model validation details (www.nic40.org).

Appendix 4 Definition of the 29 indicators

Indicators	Definition
Human capital	
Skilled labor*	Whether skilled labor is readily available
Employee training*	Whether employee training is a high propriety in companies
Literacy rate	Adult (over 15 years) literacy rate as a percentage of population
Higher education enrollment	Percentage of population that has attained at least tertiary education
Pupil-teacher ratio	Ratio of teaching staff to students
Internet subscribers	Number of Internet users per 1000 people
Public expenditure on education	Total public expenditure on education (percentage of GDP)
Market capital	
Corporate tax encouragement*	Whether corporate taxes encourage entrepreneurial activity
Cross-border venture*	Whether international transactions can be freely negotiated with foreign partners
Openness to foreign culture*	Whether the national culture is open to foreign culture ideas
Attitudes toward globalization*	Whether attitudes toward globalization are generally positive in a given society
Transparency*	Whether transparency of government policy is satisfactory
Country image*	Whether the image abroad of a given country encourages business development

(continued)

Appendix 4 (continued)

Indicators	Definition
Exports of goods	Exports of goods (percentage of GDP)
Process capital	
Business competition environment*	Whether competition legislation is efficient in preventing unfair competition
Government efficiency*	Whether government bureaucracy hinders business activity
Intellectual property rights protection*	Whether intellectual property rights are adequately enforced
Capital availability*	Whether cost of capital encourages business development
Computers in use per capita	Number of computers per 1000 people
Convenience of establishing new firms*	Whether creation of firms is supported by legislation
Mobile phone subscribers	Number of subscribers per 1000 inhabitants
Renewal capital	
Business R&D spending	Business expenditure on R&D (per capita)
Basic research*	Whether basic research enhances long-term economic development
R&D spending/GDP	Total expenditure on R&D (percentage of GDP)
R&D researchers	Total R&D personnel nationwide per capita (full-time work equivalent per 1000 people)
Cooperation between universities and enterprises*	Whether knowledge transfer is highly developed between universities and companies
Scientific articles	Scientific articles published by origin of author (per capita)
Patents per capita (USPTO + EPO)	USPTO and EPO total patents granted (per capita)
Financial capital	
GDP per capita (PPP) (from IMF)	Gross domestic product based on purchasing power parity (PPP) per capita

Appendix 5 48 Countries by Country Cluster and by Continent

This booklet series covers a total of 48 countries in 11 country clusters. China, Brazil, and South Africa are grouped twice to maintain cluster completeness. China is in both Greater China region and BRICKS region, Brazil is in both BRICKS region and Latin American region, and South Africa is in both BRICKS region and in cluster 5.

Eleven Country Clusters

1. Portugal, Italy, Greece, Spain
2. Greater China region (China, Hong Kong, Singapore, Taiwan)
3. Brazil, Russia, India, China, Korea, South Africa (BRICKS)

4. Indonesia, Malaysia, the Philippines, Thailand
5. Israel, Jordan, Turkey, South Africa
6. Bulgaria, Czech Republic, Hungary, Romania, Poland
7. Austria, Belgium, the Netherlands, Switzerland
8. France, Germany, Ireland, United Kingdom
9. Argentina, Brazil, Chile, Colombia, Mexico, Venezuela
10. Australia, Canada, Japan, New Zealand, USA
11. Denmark, Finland, Iceland, Norway, Sweden

48 Countries by Continent

Europe—24 countries

Nordic Europe—Denmark, Finland, Iceland, Norway, and Sweden,

Western Europe—Austria, Belgium, France, Germany, Ireland, the Netherlands, Switzerland, and UK

Southern Europe—Greece, Italy, Portugal, and Spain

Eastern Europe—Russia

East Central Europe—Bulgaria, Czech Republic, Hungary, Poland, Romania, and Turkey

Americas—8 countries

North America—Canada, USA

Latin America—Argentina, Brazil, Chile, Colombia, Mexico, Venezuela

Australia—2 countries

Australia and New Zealand

Asia—13 countries

East Asia—China, Hong Kong, Japan, Korea, Taiwan

West Asia—Israel, Jordan

Southeast Asia—Indonesia, Malaysia, the Philippines, Singapore, Thailand

South Asia—India

Africa—1 country

South Africa

Appendix 6 National intellectual capital scores and ranking for 48 countries (2005–2010)

2005–2010	Human capital		Market capital		Process capital		Renewal capital		Financial capital		Overall NIC	
Mean	6.420		5.409		5.345		3.658		9.115		29.946	
SD	1.217		1.085		1.454		2.056		0.690		5.840	
Country	Score	Ranking	Score	Ranking	Score	Ranking	Score	Ranking	Score	Ranking	Score	Ranking
Argentina	5.370	36	3.340	47	3.038	47	1.413	45	8.755	34	21.916	45
Australia	7.205	15	6.199	12	6.938	9	4.927	16	9.695	11	34.964	13
Austria	7.229	14	6.203	11	6.982	8	5.296	12	9.708	10	35.418	11
Belgium	7.666	8	5.519	23	5.924	21	4.706	17	9.635	14	33.450	19
Brazil	4.757	44	4.589	38	3.076	46	1.753	36	8.481	40	22.656	42
Bulgaria	5.493	34	4.929	31	4.054	36	1.589	41	8.646	37	24.710	35
Canada	7.855	6	6.137	13	6.464	15	5.064	14	9.710	9	35.229	12
Chile	5.414	35	6.663	5	5.163	28	1.710	37	8.784	32	27.734	30
China	4.942	39	5.184	28	3.746	40	2.099	33	7.967	44	23.938	40
Colombia	4.941	40	4.677	36	3.951	37	1.435	44	8.348	42	23.352	41
Czech Republic	5.995	30	5.720	18	5.378	25	2.905	24	9.271	27	29.269	25
Denmark	8.545	1	6.662	6	7.642	1	6.587	6	9.665	13	39.101	3
Finland	7.601	9	6.232	10	7.399	3	7.755	3	9.606	17	33.593	4
France	6.956	20	4.141	45	5.730	22	4.500	20	9.585	18	30.912	23
Germany	6.761	23	5.630	19	6.489	14	5.796	9	9.612	16	34.287	17
Greece	5.986	31	4.216	43	4.113	35	2.168	32	9.440	23	25.923	31
Hong Kong	6.564	25	8.090	2	7.209	5	3.277	23	9.794	4	34.934	14
Hungary	6.674	24	4.836	32	5.095	29	2.301	29	9.046	29	27.952	29
Iceland	8.357	3	5.614	21	6.561	12	5.931	8	9.713	8	36.154	8
India	3.926	48	5.620	20	3.556	42	1.954	34	7.285	48	22.340	44
Indonesia	4.062	47	4.832	33	3.257	44	1.519	42	7.594	46	21.263	47

(continued)

Appendix 6 (continued)

2005–2010	Human capital		Market capital		Process capital		Renewal capital		Financial capital		Overall NIC	
Ireland	7.145	16	6.816	3	6.711	11	4.101	21	9.768	5	34.541	16
Israel	8.040	4	5.954	16	6.454	17	7.091	4	9.414	24	36.953	6
Italy	6.471	28	4.155	44	5.184	27	2.745	26	9.479	22	28.035	28
Japan	7.581	10	4.820	34	6.209	18	6.473	7	9.576	19	34.659	15
Jordan	5.345	37	5.341	26	4.411	32	2.212	30	7.865	45	25.174	33
Korea	6.964	19	5.123	29	5.282	26	4.595	19	9.383	25	31.348	22
Malaysia	6.538	26	6.410	9	5.567	23	2.186	31	8.740	35	29.442	24
Mexico	4.871	42	4.491	39	3.226	45	1.237	48	8.781	33	22.606	43
Netherlands	7.312	12	6.761	4	6.879	10	5.163	13	9.739	7	35.855	10
New Zealand	7.123	18	5.999	15	6.460	16	3.701	22	9.380	26	32.662	21
Norway	7.936	5	6.055	14	7.008	7	4.961	15	9.987	1	35.947	9
Philippines	4.772	43	4.464	40	3.317	43	1.268	47	7.481	47	21.302	46
Poland	6.393	29	4.246	41	3.895	38	1.819	35	8.945	30	25.299	32
Portugal	6.882	22	4.801	35	5.470	24	2.467	27	9.224	28	28.844	26
Romania	5.829	32	4.219	42	3.806	39	1.685	38	8.594	39	24.133	39
Russia	5.530	33	3.860	46	3.588	41	2.349	28	8.822	31	24.150	38
Singapore	7.144	17	8.306	1	7.506	2	5.416	11	9.961	2	38.334	5
South Africa	4.708	45	4.976	30	4.472	31	1.677	39	8.472	41	24.306	37
Spain	6.524	27	4.649	37	5.033	30	2.787	25	9.484	21	28.476	27
Sweden	8.385	2	6.563	7	7.174	6	7.780	2	9.673	12	39.575	1
Switzerland	7.575	11	6.481	8	7.374	4	7.988	1	9.753	6	39.172	2
Taiwan	7.244	13	5.823	17	6.026	20	5.425	10	9.526	20	34.044	18
Thailand	5.310	38	5.524	22	4.365	33	1.436	43	8.304	43	24.938	34
Turkey	4.703	46	5.227	27	4.167	34	1.666	40	8.690	36	24.453	36
United Kingdom	6.939	21	5.440	24	6.090	19	4.697	18	9.626	15	32.793	20
USA	7.695	7	5.348	25	6.558	13	6.715	5	9.886	3	36.202	7
Venezuela	4.884	41	2.761	48	2.545	48	1.277	46	8.625	38	20.092	48

Appendix 7 Country Profile: Additional Statistics

International Reserves in Millions of US$

Country	2005	2006	2007	2008	2009	2010
Greece	816	613	584	385	285	1,479
Italy	27,456	25,342	27,121	29,469	35,752	45,172
Portugal	4,539	2,245	2,089	1,569	1,213	2,698
Spain	10,075	10,016	10,660	12,030	12,529	18,148

Source: Global finance http://www.gfmag.com/

Real GDP Growth, % Change

Country	2005	2006	2007	2008	2009	2010
Greece	2.3	5.2	4.3	1.0	−2.3	−4.4
Italy	0.7	2.0	1.5	−1.3	−5.2	1.3
Portugal	0.8	1.4	2.4	0.0	−2.5	1.3
Spain	3.6	4.0	3.6	0.9	−3.7	−0.1

Source: World economic outlook database http://www.imf.org/external/ns/cs.aspx?id=28

GDP Per Capita, PPP (Current International Dollar)

Country	2005	2006	2007	2008	2009	2010
Greece	25,076	27,174	29,106	29,978	29,571	28,434
Italy	27,944	29,297	30,410	30,412	28,883	29,392
Portugal	20,639	21,536	22,635	23,094	22,708	23,223
Spain	27,509	29,060	30,484	30,858	29,625	29,742

Source: World economic outlook database http://www.imf.org/external/ns/cs.aspx?id=28

GDP (PPP), Share of World Total (%)

Country	2005	2006	2007	2008	2009	2010
Greece	0.49	0.49	0.49	0.48	0.47	0.43
Italy	2.88	2.80	2.70	2.59	2.48	2.39
Portugal	0.38	0.37	0.36	0.35	0.35	0.33
Spain	2.09	2.07	2.03	2.00	1.94	1.85

Source: World economic outlook database http://www.imf.org/external/ns/cs.aspx?id=28

Gross Domestic Expenditure on R&D as a % of GDP

Country	2005	2006	2007	2008	2009	2010
Greece	0.62	0.49	0.57	0.57	0.58	0.58
Italy	1.13	1.1	1.09	1.13	1.19	1.27
Portugal	0.74	0.8	0.81	1.18	1.51	1.66
Spain	1.05	1.12	1.12	1.2	1.35	1.38

Source: IMD world competitiveness yearbook 2006–2011

Household Saving Rates, Percent of Disposable Household Income

Country	2005	2006	2007	2008	2009	2010
Greece	–	–	–	–	–	–
Italy	9.9	9.1	8.4	8.0	6.5	5.0
Portugal	10.0	8.0	7.0	7.1	10.9	9.7
Spain	10.8	10.2	10.4	13.5	18.5	13.9

Source: Global finance http://www.gfmag.com/

Glossary

APEC Asia-Pacific Economic Cooperation
Basel II **Basel II** is the second of the Basel Accords, which are recommendations on banking laws and regulations issued by the Basel Committee on Banking Supervision. Basel II, initially published in June 2004, was intended to create an international standard for banking regulators to control how much capital banks need to put aside to guard against the types of financial and operational risks banks face
BBC British Broadcasting Corporation
BIS Bank for International Settlements
BRICKS Brazil, Russia, India, China, Korea, South Africa
Cajas Unregulated Private Banks in Spain
CIA The US Central Intelligence Agency
CPI Consumer Price inflation
ECB Euro-convertible Bond
EFSF European Financial Stability Facility
EFSM European Financial Stabilization Mechanism
EMU Economic and Monetary Union
EPO European Patent Office
ESM European Stabilization Mechanism
EU European Union
European Council One of the EU's Regulatory Bodies
Eurostat The EU's Statistical Office
FC Financial Capital
FDI Foreign Direct Investment
FROB Fondo de reestructuración ordenada bancaria (FROB) is a banking bailout and reconstruction program initiated by the Spanish government in June 2009
FSF Financial Stability Forum
GCI Global Competitiveness Index (published by World Economic Forum)
GDP Gross Domestic Product
GDP Per capita (ppp) Gross domestic product per capita adjusted by purchasing power parity

C.Y.-Y. Lin et al., *National Intellectual Capital and the Financial Crisis in Greece, Italy, Portugal, and Spain*, SpringerBriefs in Economics 7,
DOI 10.1007/978-1-4614-5990-3,

IMD International Institute for Management Development in Switzerland
IMF International Monetary Fund
HC Human Capital
IP Intellectual Property
Long-term NIC Long-term-oriented national intellectual capital, represented by human capital and renewal capital
MC Market Capital
MTFS Greece's Medium-term Fiscal Strategy
NIC National intellectual Capital
OECD Organization for Economic Cooperation and Development
PC Process Capital
PIIGS Portugal, Ireland, Italy, Greece, and Spain
PPP Purchasing Power Parity
R&D Research and Development
RC Renewal Capital
SAR Special Administrative Region (Hong Kong SAR)
SD Standard Deviation
Short-term NIC Short-term-oriented national intellectual capital, represented by market capital and process capital
SMEs Small- and Medium-sized Enterprises
Stability and Growth Pact The pact required that countries maintain a government budget deficit of no more than 3 % of the GDP and a country's public debt cannot exceed 60 % of the GDP
Three time periods For the convenience of comparing ranking changes, the six-year data were separated into three time periods, namely, 2005–2006, 2007–2008, and 2009–2010. Two years each represents "before, during, and after" the 2008 global financial crisis. Even though Euro zone is still in sovereign debt trouble as of mid-2012, the time partition is based on the financial crisis progression in the epicenter. The signs of financial trouble appeared in 2007, and the crisis was declared over in the third quarter of 2009
UNCTAD United Nations Conference on Trade and Development
USPTO United States Patent and Trademark Office
VAT Value-added Tax

References

Abboushi S (n.d.) Analysis and outlook of the Greek Financial Crisis. Retrieved from http://www.jgbm.org/page/9%20Suhail%20Abboushi.pdf

Al Jazeera and Agencies (March 8, 2012) Eurozone approves second Greek bailout, Al Jazeera and Agencies. Retrieved from http://www.aljazeera.com/news/europe/2012/02/201222132429323747.html

Altman RC (January 2, 2009) The great crash, 2008: a geopolitical setback for the west, Foreign Affair, the Council on Foreign Relations. Retrieved from http://www.foreignaffairs.com/articles/63714/roger-c-altman/the-great-crash-2008

Banco De Portugal (December, 2011) Outlook for the Portuguese economy: 2011–2013, Economic Bulletin, Winter 2011. Retrieved from http://www.bportugal.pt/en-US/EstudosEconomicos/Publicacoes/BoletimEconomico/Publications/projecoes_e.pdf

Barnes S (2010) OECD Economic Surveys: Euro Area 2010, Resolving and avoiding unsustainable imbalances, Economics Department working paper No. 827. Retrieved from http://www.keepeek.com/Digital-Asset-Management/oecd/economics/oecd-economic-surveys-euro-area-2010/resolving-and-avoiding-unsustainable-imbalances_eco_surveys-euz-2010-5-en

BBC (April 24, 2009) Spain's jobless rate soars to 17 %, BBC News. Retrieved from http://news.bbc.co.uk/2/hi/business/8016364.stm

BBC (May 16, 2011) Portugal's 78bn euro bail-out is formally approved, BBC News. Retrieved from http://www.bbc.co.uk/news/business-13408497

BBC (February 13, 2012a) Timeline: the unfolding Eurozone crisis, BBC News. Retrieved from http://www.bbc.co.uk/news/business-13856580

BBC (March 2, 2012b) Q&A: Greek debt crisis, BBC News. Retrieved from http://www.bbc.co.uk/news/business-13798000

Blavoukos S, Pagoulatos G (2008) The limits of EMU conditionality: fiscal adjustment in Southern Europe. Journal of Public Policy 28:229–253

Bryant L (September 8, 2011) Growth expected to slow in world's largest economies, Voice of America News. Retrieved from http://www.voanews.com/english/news/europe/OECD--Economic-Slowdown-Likely-to-Continue-129458543.html

Castle S (March 14, 2012) With details settled, a 2nd Greek bailout is formally approved, The New York Times. Retrieved from http://www.nytimes.com/2012/03/15/business/global/greece-gets-formal-approval-for-second-bailout.html

Cecchetti SG, Mohanty MS, Zampolli F (2011) The real effects of debt. Bank for International Settlements (BIS), prepared for the "Achieving Maximum Long-run Growth" symposium sponsored by the Federal Reserve Bank of Kansas City, Jackson Hole, Wyoming, 25–27 August 2011

Central Intelligence Agency (CIA) (2011 and 2012) The world factbook, U.S.A. Central Intelligence Agency. Retrieved from https://www.cia.gov/library/publications/the-world-factbook/geos/ch.html

C.Y.-Y. Lin et al., *National Intellectual Capital and the Financial Crisis in Greece, Italy, Portugal, and Spain*, SpringerBriefs in Economics 7,
DOI 10.1007/978-1-4614-5990-3, © The Author(s) 2013

Dilje.com. (November 28, 2008). Italy approves stimulus plan, analysts skeptical. Retrieved from http://dalje.com/en-economy/italy-approves-stimulus-plan-analysts-sceptical/208950
Dimireva I (October 28, 2009) Greece: economy overview, EUbusiness. Retrieved from http://www.eubusiness.com/europe/greece
Edvinsson L, Malone MS (1997) Intellectual capital: realizing your company's true value by finding its hidden brainpower. Harper Business, New York
Economy Watch (June 30, 2010a) Italy economic stimulus package. Retrieved from http://www.economywatch.com/economic-stimulus-package/italy.html
Economy Watch (June 30, 2010b) Portugal economic stimulus package. Retrieved from http://www.economywatch.com/economic-stimulus-package/portugal.html
Euro Challenge (2012) An overview of Portugal's economy. Retrieved from http://www.euro-challenge.org/doc/Portugal.pdf
Eurofound (2010) Italy: annual review – 2010, Eurofound. Retrieved from http://www.eurofound.europa.eu/comparative/tn1105040s/it1105049q.htm
Europa (November 26, 2008). The commission launches a major recovery plan for growth and jobs, to boost demand and restore confidence in the European economy. Retrieved from http://europa.eu/rapid/pressReleasesAction.do?reference=IP/08/1771&format=HTML&aged=0&language=EN&guiLanguage=en
European Commission. (2010). European innovation, Progress report 2009. Pro Inno Europe Paper No.17. Retrieved from http://www.insme.org/files/4221
Ezrati M (April 24, 2012) Europe's bleak prospects: austerity obsession creates a vicious cycle. Treasury and Risk. Retrieved from http://www.treasuryandrisk.com/2012/04/24/ezratis-economic-outlook-europes-bleak-prospects
Fackler M (October 23, 2008) Financial crisis spreads to emerging nations. The New York Times. Retrieved from http://www.nytimes.com/2008/10/24/business/worldbusiness/24won.html
First Post (August 8, 2011) PIIGS, first post. Retrieved from http://www.firstpost.com/topics/piigs-56382.html
Global Finance (2012) Greece country report, Global Finance. Retrieved from http://www.gfmag.com/gdp-data-country-reports/266-greece-gdp-country-report.html
Gurria A (2011a) Economic Survey of Greece 2011, OECD Secretary-General. Retrieved from http://www.oecd.org/document/28/0,3746,en_2649_34117_48421852_1_1_1_1,00.html
Gurria A (2011b) Economic survey of Italy 2011, OECD Secretary-General. Retrieved from http://www.oecd.org/document/8/0,3746,en_2649_34327_47725832_1_1_1_1,00.html
Harrington C (April 2011) The Spanish financial crisis. International Finance and Development, University of Iowa. Retrieved from http://blogs.law.uiowa.edu/ebook/content/spanish-financial-crisis
Heritage Foundation (2011) Spain. 2012 index of economic freedom. Retrieved in October 2011 from http://www.heritage.org/index/country/Spain
Hugh E (May 9, 2010) Spain emerges from recession? Spain economy watch. Retrieved from http://spaineconomy.blogspot.com/2010/05/spain-emerges-from-recession.html
IMF (November 19, 2008) Italy: 2008 Article IV consultation – Concluding statement of the mission. International Monetary Fund. Retrieved from http://www.imf.org/external/np/ms/2008/111908.htm
Inflation.eu (2011) Inflation Greece 2011, Worldwide Inflation data. Retrieved from http://www.inflation.eu/inflation-rates/greece/historic-inflation/cpi-inflation-greece-2011.aspx
Kehoe T (2010) The current financial crisis in Spain: what should we learn from the great depressions of the twentieth century? University of Minnesota and Federal Reserve Bank of Minneapolis, 18th Barcelona Lecture, Barcelona G.S.E. and Banc Sabadell, February 2010.
Knight L (2011) What's the matter with Italy? BBC News. Retrieved from http://www.bbc.co.uk/news/business-15429057
Krugman P (February 5, 2010) The Spanish tragedy. New York Times. Retrieved from http://krugman.blogs.nytimes.com/2010/02/05/the-spanish-tragedy/
Lin CYY (2011) Leveraging intellectual capital for future competitiveness in Finland, Presentation at Tekes, Helsinki on November 10, 2011

Lin CYY, Edvinsson L (2011) National intellectual capital: a comparison of 40 countries. Springer Publishing Co., New York

Lynn M (2011) Bust: Greece, the Euro and the Sovereign debt crisis. Bloomberg Press, Hobeken. ISBN 978-0-470-97611-1

Mallet V (March 31, 2011) Merger of 4 Spanish savings banks collapses, Financial Times. Retrieved from http://www.cnbc.com/id/42353105/Merger_of_4_Spanish_Savings_Banks_Collapses

Minder R (May 3, 2011) Portugal agrees to a $116 billion bailout, The New York Times, http://www.nytimes.com/2011/05/04/business/global/04portugal.html

Moloney N (2010) EU financial market regulation after the global financial crisis: "More Europe" or more risks? Common Market Law Review 47:1317–1383

Nanto DK (October 2, 2009) The global financial crisis: Analysis and policy implications. CRS Report for Congress, Congressional Research Service 7–5700, RL34742

Nelson RM, Belkin P, Mix DE (2011) Greece's debt crisis: overview, policy responses, and implications. Retrieved from http://www.fas.org/sgp/crs/row/R41167.pdf

Norris F (March 10, 2011) Crisis is over, but where's the fix? The New York Times. Retrieved from http://www.nytimes.com/2011/03/11/business/economy/11norris.html?pagewanted=1

OECD (June 2009) Policy responses to economic crisis: Investing in innovation for long-term growth. OECD. Retrieved from http://www.oecd.org/dataoecd/59/45/42983414.pdf

OECD (2010) 2010 Economic Surveys Portugal, OECD. Retrieved from http://www.oecd.org/dataoecd/34/20/46062172.pdf

OECD (August 2011a) 2011 Economic Surveys Greece, OECD. Retrieved from http://www.oecd.org/dataoecd/21/63/48462569.pdf

OECD (May 2011b). 2011 Economic Surveys Italy, OECD. Retrieved from http://www.oecd.org/dataoecd/58/29/47769054.pdf

OECD (n.d) Italy fiscal stimulus package and strategies to reduce fiscal deficit, OECD. Retrieved from http://www.oecd.org/dataoecd/42/30/47741933.pdf

Pagoulatos G (2010) The Greek economy and the potential for green development. Retrieved from http://library.fes.de/pdf-files/id/ipa/07097.pdf

Quirico RD (2010) Italy and the global economic crisis. Bulletin of Italian Politics 2(2):3–19. Retrieved from http://www.gla.ac.uk/media/media_191024_en.pdf

Reinhart CM, Rogoff KS (2009) This time is different: eight centuries of financial folly. Princeton University Press, Princeton. ISBN 9780691142166

Reavis C (2009) The global financial crisis of 2008–2009: the role of greed, fear and oligarchs, MIT Sloan Management Case #09-093, published 2009.07.22

Reguly E (April 6, 2011) European financial crisis: It's not over, Eupedia. Retrieved from http://www.eupedia.com/forum/showthread.php?26399-European-Financial-Crisis-It-s-not-over

Report Linker (February 27, 2012) Portugal: economy statistics and industry reports, Report Linker. Retrieved from http://www.reportlinker.com/r0818/Portugal-industry-reports.html

Rogoff K (2011) in Javetski B and Koller T (ed) Understanding the second great contraction: An interview with Kenneth Rogoff. McKinsey Quarterly. Retrieved from https://www.mckinseyquarterly.com/Economic_Studies/Productivity_Performance/Understanding_the_Second_Great_Contraction_An_interview_with_Kenneth_Rogoff_2871

RTE (June 10, 2012) Eurozone agrees deal to lend Spain up to €100bn as Economy Minister Luis de Guindos confirms request for financial assistance, RTE News. Retrieved from http://www.rte.ie/news/2012/0609/spain-bailout.html

Samuelson RJ (April 23, 2012) The pain in Spain could hit worldwide economy, The Washington Post. Retrieved from http://www.washingtonpost.com/opinions/the-pain-in-spain-could-hit-worldwide-economy/2012/04/22/gIQAQCfiaT_story.html

Schwab K (ed) (2011) The global competitiveness report 2011–2012. World Economic Forum, Geneva

Sorsa P, Giorno C, Koutsogeorgopoulou V (July 2009) Economic survey of Greece, 2009 Policy Brief, OECD. Retrieved from http://www.oecd.org/dataoecd/5/61/43284926.pdf

Stratfor (April 28, 2009) The financial crisis in Spain, Stratfor Global Intelligence. Retrieved from http://www.stratfor.com/memberships/136936/analysis/20090428_financial_crisis_spain

The Economist (May 15, 2008) Spanish steps: a simple way of curbing banks' greed, The Economist Special Report: International banking. Retrieved from http://www.economist.com/node/11325484?story_id=11325484

The Economist (April 13, 2012) European economy guide: polarized prospects, The economist online. Retrieved from http://www.economist.com/blogs/dailychart/2010/12/europes_economies

The New York Times (March 16, 2012) Greece, The New York Times. Retrieved from http://topics.nytimes.com/top/news/international/countriesandterritories/greece/index.html

Tirone M (September 26, 2011) Chart: Who owns our government debt? Wealth Wire. Retrieved from http://www.wealthwire.com/news/economy/1923

Wang Y (March 11, 2009) Old obstacles, new crisis hits Italy's lagging economy, Xinhua Net. Retrieved from http://news.xinhuanet.com/english/2009-03/11/content_10990277.htm

Wolls D, Dilorenzo S (June 10, 2012) Europe bailout of Spain could cost $125 billion. Yahoo News. Retrieved from http://news.yahoo.com/europe-bailout-spain-could-cost-125-billion-190614499--finance.html

Author Index

C.Y.-Y. Lin et al., *National Intellectual Capital and the Financial Crisis in Greece, Italy, Portugal, and Spain*, SpringerBriefs in Economics 7,
DOI 10.1007/978-1-4614-5990-3,

Subject Index

C.Y.-Y. Lin et al., *National Intellectual Capital and the Financial Crisis in Greece, Italy, Portugal, and Spain*, SpringerBriefs in Economics 7,
DOI 10.1007/978-1-4614-5990-3,

MIX
Papier aus verantwortungsvollen Quellen
Paper from responsible sources
FSC® C105338

If you have any concerns about our products,
you can contact us on
ProductSafety@springernature.com

In case Publisher is established outside the EU,
the EU authorized representative is:
Springer Nature Customer Service Center GmbH
Europaplatz 3, 69115 Heidelberg, Germany

Printed by Libri Plureos GmbH
in Hamburg, Germany